T0237290

SpaceX

Starship To Mars – The First 20 Years

Second Edition

Erik Seedhouse

SpaceX

Starship To Mars – The First 20 Years

Second Edition

 Springer

Published in association with
Praxis Publishing
Chichester, UK

Erik Seedhouse
Applied Aviation Sciences
Embry-Riddle Aeronautical University
Daytona Beach, FL, USA

SPRINGER-PRAXIS BOOKS IN SPACE EXPLORATION

Space Exploration
ISSN 2731-5401   ISSN 2731-541X   (eBook)
Springer Praxis Books
ISBN 978-3-030-99180-7   ISBN 978-3-030-99181-4   (eBook)
https://doi.org/10.1007/978-3-030-99181-4

© Springer Nature Switzerland AG 2013, 2022
This work is subject to copyright. All rights are reserved by the Publisher, whether the whole or part of the material is concerned, specifically the rights of translation, reprinting, reuse of illustrations, recitation, broadcasting, reproduction on microfilms or in any other physical way, and transmission or information storage and retrieval, electronic adaptation, computer software, or by similar or dissimilar methodology now known or hereafter developed.
The use of general descriptive names, registered names, trademarks, service marks, etc. in this publication does not imply, even in the absence of a specific statement, that such names are exempt from the relevant protective laws and regulations and therefore free for general use.
The publisher, the authors and the editors are safe to assume that the advice and information in this book are believed to be true and accurate at the date of publication. Neither the publisher nor the authors or the editors give a warranty, expressed or implied, with respect to the material contained herein or for any errors or omissions that may have been made. The publisher remains neutral with regard to jurisdictional claims in published maps and institutional affiliations.

Project Editor: Michael D. Shayler

This Praxis imprint is published by the registered company Springer Nature Switzerland AG
The registered company address is: Gewerbestrasse 11, 6330 Cham, Switzerland

Contents

DEDICATION

To Lava

Deeply loved, completely spoiled and missed beyond words.

Lava was a Nebelung, one of the more social and intelligent breeds. He used to walk with our dogs and became something of a minor celebrity where we live because people used to stop their cars and ask us if Lava was really a cat! Writing can be a lonely pursuit with long hours spent at a keyboard, and having written 30 books, that could have added up to a lot of time alone. But sitting next to me for much of my writing, purring like a tractor, was Lava, who very sadly crossed over the Rainbow Bridge shortly after this book was completed.

Acknowledgements

In writing this book, the author has been fortunate to have had five reviewers who made such positive comments concerning the content of this publication. He is also grateful to Hannah Kaufman at Springer and to Clive Horwood and his team at Praxis for guiding this book through the publication process. The author also gratefully acknowledges all those who gave permission to use many of the images in this book. The author also expresses his deep appreciation to Mike Shayler, whose attention to detail and patience greatly facilitated the publication of this book, and to Project Manager, Ms. A. Meenahkumary and her team for producing the final proof.

About the Author

Erik Seedhouse works as a professor in Space Operations at Embry-Riddle Aeronautical University, located just up the road from the Space Coast in Florida. He lives in Flagler Beach, which means he can often watch SpaceX launches from his lanai. After completing his first degree in Sports Science at Northumbria University in the UK, Erik joined the 2nd Battalion the Parachute Regiment, the world's most elite airborne regiment. During his time in the 'Paras', Erik spent several months involved in various counternarcotics operations in Belize (anyone who has seen the series 'Narcos' will be familiar with what these operations entailed). Later, he spent several months learning the intricacies of desert warfare on the Akamas Range in Cyprus. He made more than 30 jumps from a Hercules C130 aircraft, performed more than 200 abseils from a helicopter and fired more light anti-tank weapons than he cares to remember!

Upon returning to the comparatively mundane world of academia, the author embarked upon a master's degree at Sheffield University, also in the UK. He supported his studies by winning prize money in 100km running races. After placing third in the World 100km Championships in 1992 and setting the North American 100km record, the author turned to ultra-distance triathlon, winning the World Endurance Triathlon Championships in 1995 and 1996. For good measure, he also won the inaugural World Double Ironman Championships in 1995 and the infamous Decatriathlon, an event requiring competitors to swim 38km, cycle 1,800km, and run 422km. Non-stop!

Returning to academia in 1996, Erik pursued his Ph.D. at the German Space Agency's Institute for Space Medicine through a grant from the European Space Agency – thanks ESA! While conducting his Ph.D. studies he still found time to win Ultraman Hawai'i and the European Ultraman Championships, as well as completing the Race Across America bike race. Due to his success as the world's leading ultra-distance triathlete, Erik was featured in dozens of magazines and television interviews. In 1997, GQ magazine nominated him as the 'Fittest Man in the World'.

In 1999, Erik decided it was time to get a real job. He retired from being a professional triathlete and started post-doctoral studies at Simon Fraser University's School of Kinesiology in sunny Vancouver, Canada. In 2005, the author worked as an astronaut training consultant for Bigelow Aerospace and wrote 'Tourists in Space', a training manual for spaceflight participants. In 2009, he was one of the final group of candidates in the Canadian Space Agency's Astronaut Recruitment Campaign. He came close but not close enough. Between 2008 and 2012, he served as director of the Canadian Forces manned centrifuge and hypobaric chamber operations. In addition to his work as a professor, Erik works as a manned spaceflight consultant, professional speaker, triathlon coach and author. He has also worked as a film consultant to Hollywood on such productions as '*Into the Unknown*' featuring Mark Strong. He is an instructor for the International Institute of Astronautical Sciences and completed his suborbital scientist astronaut training in 2016. At the time of writing, Erik has written 30 books. When not writing, he can be found climbing big mountains or spending time with his wife Alice on the Big Island of Hawai'i, or in Sandefjord, Norway.

Prologue

"We want the exciting things we see in sci-fi, in like sci-fi movies, books – we want them to come true one day."

"Consciousness is a very rare and precious thing, we should take whatever steps we can to preserve the light of consciousness. We should do our very best to become a multi-planet species, extend consciousness beyond Earth, and we should do it now."

Elon Musk. The Rise of Elon Musk's Engineering Masterpiece

SpaceX is like no other rocket company. It was started by someone who knew nothing about rockets but who went on to become the chief technology officer of a company whose vehicles have turned the launch industry on its head in a very short span of time. Elon Musk does not have an aerospace degree; he is self-taught, meaning he read a lot of books and talked to a lot of people. Shortly after SpaceX was launched, Apollo astronauts Neil Armstrong and Gene Cernan – two of Musk's heroes who inspired the Mars Messiah to start his company in the first place – testified against commercial spaceflight, saying the way SpaceX was going about the business of building launch vehicles was all wrong. Yet despite the nay-sayers, despite the trauma of 2008 when Tesla was almost bankrupt and the American economy had tanked in the worst recession since the Great Depression, despite his first three rockets failing, despite crisis after crisis that would have crushed anyone else, Musk prevailed. When SpaceX finally managed to launch to orbit after three unsuccessful flights, NASA called Musk and told him the agency had awarded the company one and a half billion dollars. That launch was a supreme achievement, because at the time only four entities had launched a rocket into orbit – the United States, Russia, China… and SpaceX. Since then, SpaceX has racked up success after success as the company grabbed the headlines with landing of a first stage, the launch of a Tesla on the Falcon Heavy, and of course, the development of the Starship.

In this book we examine the achievements of perhaps the most remarkable company in human history: one that may just transform the human race into becoming an interplanetary species. We discuss the development of the Dragon, the first privately developed spacecraft to visit the International Space Station (ISS). Not only was Dragon the first privately developed spacecraft to successfully return from Earth orbit, it is also the only reusable spacecraft in operation today. It also happens to be another piece of the puzzle in Elon Musk's goal of making humanity a space-faring civilization. Then there is the family of launch vehicles developed, tested and flown in rapid succession. How? Well, we will see how SpaceX has applied modern manufacturing techniques, such as friction stir welding and modern CAD and production data management techniques, to building its rockets. SpaceX developed its Falcon 1, 9, Falcon Heavy and Starship so rapidly by reusing many components and applying cutting edge design and manufacturing strategies. Not satisfied with business as usual, SpaceX does not rely on decades-old space-proven products, or even the veteran aerospace testing firms; instead, it builds new components and tests them in-house.

'*SpaceX – The First 20 Years*', is an account of commercial spaceflight's most successful start-up. It describes the extraordinary feats of engineering and human achievement that have placed SpaceX at the forefront of the launch industry and positioned it as the most likely candidate for transporting humans to Mars. Since its inception in 2002, SpaceX has changed the space launch paradigm by developing a family of launch vehicles that have steadily reduced the cost and increased the reliability of space access by a factor of ten. Coupled with the newly emerging market for private and commercial space transport, this new model has re-ignited our efforts to explore and develop space. Here, in this book, is a portrait of perhaps one of the most spectacular aviation triumphs of the 21st century.

1

Elon Musk: Mars Messiah

"I think that when I say multi-planet species, that's what we really want to be. It's not like still being a single-planet species by moving planets. It's really being a multi-planet species and having civilization and life as we know it extend beyond Earth to the rest of the Solar System, and ultimately to other star systems. That's the future, that's exciting and inspiring, and I think that… you need things like that to be glad to wake up in the morning. Life can't be just about solving problems. There have to be things that are exciting and inspiring that make you glad to be alive."

Elon Musk, Code Conference, June 2016

Figure 1.0. Credit NASA. Illustration by Dave Mosher

© Springer Nature Switzerland AG 2022
E. Seedhouse, *SpaceX*, Springer Praxis Books,
https://doi.org/10.1007/978-3-030-99181-4_1

The goal of creating a multi-planet species is a bold one; and it would probably be judged too bold if promoted by anyone other than Elon Musk (Figure 1.0). But, thanks to a string of successes for SpaceX with its Falcon family of rockets, myriad cargo and crew missions to the International Space Station (ISS), and the Starship, Musk's much touted crewed mission to Mars has gained and continues to gain traction. It is a welcome change for the thousands of die-hard Mars enthusiasts who have put up with decades of Mars conferences presenting PowerPoint slides and computer-generated imagery of hypothetical crewed missions that never went anywhere. NASA's Design Reference Mission (DRM) 1.0 anyone? DRM 5.0 perhaps? Or what about Inspiration Mars or Mars One? Given the pace at which Musk's company is developing the hardware necessary to transport humans to Mars, it is not surprising that many see the Chief Rocket Designer of SpaceX as a sort of Mars Messiah.

For a long time, the 'Mars in a Decade' crowd wanted to believe the myriad keynote speakers who promoted the idea that a crewed Mars mission could be achieved by 2010, or if not by 2010 then by 2020. Books were published about how a crewed Mars mission could be achieved in 2030, or by 2040; using government money, no less. As we know, these cry wolf claims were no more than pipe dreams perpetuated by the fake news narrative of the Cigarette Smoking Men of the Mars Underground. But no more. Since Musk's multiplanetary species announcement in 2016, his company has launched one trailblazing mission after another. There was that Tuesday afternoon in February 2018 when the world's most powerful rocket arced upward high above Florida's Space Coast (Figure 1.1). There was that Sunday in November 2020 when the first operational crewed mission to the ISS launched onboard the Crew Dragon (Figure 1.2). And, of course, there are all those Starship launches (Figure 1.3), featuring the very vehicle that will transport humans to the Red Planet sometime before the end of the 2020s.

A crewed Mars mission is at the very edge of what is currently technologically possible and is a supremely risky proposition, but Musk loves risk. The naysayers, of which there have been many over the years, say he is reckless, but the gambles Musk has taken throughout, although risky, have almost always paid off. Take the example of perhaps his biggest gamble of all: the creation of SpaceX. This story is described in Ashley Vance's biography *Elon Musk: Tesla, SpaceX, and the Quest for a Fantastic Future*, which is well worth reading. Musk and a group of aerospace engineers were returning from Moscow, where they had been snubbed with their offer to buy refurbished Russian intercontinental ballistic missiles (ICBM) from ISC Kosmotras (the Russian sticker price was $8 million). During his meeting with the Russians, Musk was dismissed as a novice and was even spat on by one of the senior Russian rocket engineers. Not surprisingly, Musk was not happy, but after some feverish crunching of numbers on a spreadsheet, he announced he had a plan to build a rocket, cut launch prices by a factor of ten *and* still have a 70 percent gross margin. Accompanying Musk on the flight were founding SpaceX member and ex-Jet Propulsion Laboratory (JPL) employee Jim

Figure 1.1. The launch of the Falcon Heavy was the death knell for many of Musk's commercial spaceflight rivals[1], but more specifically its payload underscored the primary existence of SpaceX. Carried in the Falcon's upper stage was Musk's personal cherry red Tesla Roadster, piloted by a dummy astronaut listening to David Bowie's *Space Oddity*. After the launch, Musk voiced the hope that the Falcon Heavy could help NASA return to the Moon and perhaps prepare for an eventual landing on the Red Planet by the middle of the century. Credit NASA.

Cantrell, and future NASA administrator Mike Griffin, who both knew more than most about the challenges of launching rockets. They were skeptical to say the least, until they studied Musk's numbers on the spreadsheet. The spreadsheet detailed estimated performance characteristics of Musk's proposed rocket in exquisite detail. This surprised Cantrell and Griffin, because Musk had no background in aerospace engineering. But he had been studying. For months, Musk had immersed himself in the fundamentals of rocket propulsion, the theory of astrodynamics and the basics of aerothermodynamics. Having devoured as many seminal texts on these subjects as he could lay his hands on, Musk had turned himself into something of an expert and he used this expertise to redefine the launch vehicle arena and the commercial spaceflight industry.

---

[1] SpaceX priced a Falcon Heavy launch at about $90 million in 2016. This price compared to the $435 million cost of a Delta IV operated by United Launch Alliance. The main difference is that the Falcon Heavy can carry almost twice the payload of a Delta – and the fact that the Delta cannot be reused.

Figure 1.2. The SpaceX Crew-1 official crew portrait, with (from left) NASA astronauts Shannon Walker, Victor Glover, and Mike Hopkins, and JAXA (Japan Aerospace Exploration Agency) astronaut Soichi Noguchi. Credit NASA.

Yet despite the stratospheric rise of SpaceX over the past 20 years, there are still some who worry that risk-taking at the level that Musk is pursuing might be his undoing. The cynics often cite examples of other Musk companies to prove their point. For example, Musk's first company, Zip2, was bought by Compaq in 1999. Most of the $20 million Compaq paid Musk was channeled into another startup that became PayPal, which was bought by eBay in 2002. That transaction netted Musk $180 million, almost all which was used to develop SpaceX and then Tesla, a venture for which Musk put almost his entire fortune at risk to support. In 2009, for example, Musk borrowed $20 million against SpaceX to use in his Tesla startup. This was not a problem because SpaceX was flush with money, having received $1.6 billion from NASA for cargo flights. Fast forward to 2018 and Musk was still using the same borrowing strategy, but the numbers – and the risks – were higher. In late 2018, Musk owed $625 million on his Tesla stock which was worth $10 billion. The problem was that, at the time, Tesla was billions of dollars in debt, a set of circumstances which prompted suggestions that SpaceX should acquire

Figure 1.3. Launch of Starship SN15 on May 5, 2021. The flight tested several upgrades and culminated in a smooth touchdown on the landing pad at Boca Chica, Texas. The test occurred amid preparations for the push to orbit that were taking place at the Orbital Launch Site. Credit Jack Beyer, NASA.

Tesla. That sounds simple, but with Tesla struggling, managing that debt could have placed the plans for SpaceX in peril. As many remember, Tesla nearly toppled into bankruptcy, but the company has since become the most valuable and, by certain metrics, the most profitable carmaker on the planet. Once again, Musk's risk-taking paid off, reminding everyone that he is not just any CEO.

Back in 2002, Musk was just another Internet mogul starting a commercial space company, but Musk was bolder than his peers. Simply providing a suborbital trip to space like Richard Branson's *SpaceShipOne*[2] was never going to satisfy the South African native; Musk wanted to fly resupply missions with astronauts to the ISS and use that as a stepping-stone to Mars. It was a bold goal because, as any space engineer will tell you, getting to orbit is more difficult than reaching

---

[2] That rocket, and the passenger version – *SpaceShipTwo* – that makes up Richard Branson's Virgin Galactic fleet, was supposed to fly to 100 kilometers altitude, but operational flights have so far only reached 85 kilometers.

suborbital altitudes, by several orders of magnitude. In fact, it is such a challenge that only eight countries and a few private companies have reached orbit successfully. Orbital flight also happens to be very, *very* expensive, but Musk maintained that he could do it cheaper *and* turn a profit. His plan? Run his company like an Internet startup and launch a new age in space exploration along the way.

One of the most intriguing aspects about how Musk works is the fact that he works at all. By his early thirties, his Internet ventures had made his net worth about $200 million (at the time of writing he is worth more than $300 billion, making him the richest person in history). He could have retired, but chose instead to enter the riskiest, costliest, and most unforgiving business there is: launching rockets. Born in South Africa in 1971, the son of a Canadian mother and a South African father, it did not take long for Musk to demonstrate his entrepreneurial spirit. He bought his first computer at the age of ten and quickly taught himself computer programming. Two years later, he wrote code for a video game called *Blastar*, which he subsequently sold to a computer magazine for $500. Then, when he was 17, and spurred on by the prospect of avoiding compulsory service in the South African military[3], Musk moved to Canada, spending two years at Queen's University, Kingston. He had planned a career in business and worked at a Canadian bank for one summer as a college intern. After Kingston, Musk moved to the U.S., where he earned degrees in Physics and Business at the University of Pennsylvania. He had intended to begin a graduate program at Stanford in 1995, but chose instead to devote the next four years to developing Zip2, a company that enabled other companies to post content on the Internet. In February 1999, Compaq Computer Corporation bought Zip2 for $307 million, in cash. It was one of the largest cash deals in the Internet era at the time, and Musk walked away with a cool $22 million. He was only 28.

He used $10 million to start X.com, an online bank, which went online in December 1999. The following month, Musk married his first wife, Justine, whom he had met while studying in Canada. Two months later, in March 2000, X.com merged with Confinity, which had developed a service you may have heard of called PayPal. Musk increased his fortune when eBay bought PayPal for $1.5 billion in 2002, a deal that saw his net worth rocket past $100 million. By that time, he and Justine had moved to Los Angeles and had their first child. Tragically, Musk's son stopped breathing while having a nap one day, and by the time the paramedics had resuscitated him the ten-week-old infant had been without oxygen for so long that he was pronounced brain-dead. He spent three days on life support

---

[3] Musk has explained in several interviews that he does not have a problem with military service, but that he did not like what the South African military was doing in the late 1980s, especially the brutal oppression of the black majority. When he moved to Canada to avoid conscription it was against the wishes of his father, and the two rarely speak because of the younger Musk's decision.

before Musk and his wife made the agonizing decision to have it turned off. The verdict was Sudden Infant Death Syndrome.

Having had enough of the Internet, Musk searched for a new challenge and founded Space Exploration Technologies, or SpaceX. To kick-start his company he tried buying that ICBM from Russia, but having been rebuffed he decided instead to build his own. Establishing a rocket company was seen by many in the space industry as an audacious move, particularly since Musk possessed little background in the field of rocket science. He could have been forgiven if he had chosen to buy rockets from established rocket-building companies, but that would not have been Musk. Instead, he decided to build SpaceX from the ground up. His initial goal was to reduce the cost of launch services, a milestone prompted by Musk's frustration with both how much money NASA spent on the space program and how little the costs of space exploration had decreased since the end of the Apollo Program in the 1970s. Once he had solved the inefficiencies of the space program, Musk had his sights set on low-cost human travel into orbit and establishing a colony on Mars. But before he could send anyone to Mars, Musk first needed to get his rockets into orbit.

The challenges facing Musk were formidable. Between 1957 and 1966, just as the Space Age was gaining momentum, the U.S. had sent 429 rockets into orbit, a quarter of which had failed. At around the time Musk was looking to get into the rocket business, only governments had managed to harness the capital and intellectual muscle necessary to launch payloads into orbit, and building those rockets did not come cheap. The American, Russian and Chinese space programs required armies of engineers working with nearly unlimited budgets. For example, the Apollo Program employed more than 300,000 people and cost more than $150 billion in 2007 dollars, or more than three percent of the U.S. federal budget. Even the now-retired Space Shuttle required a ground crew of 50,000 and cost more than one billion dollars every time it flew. Incidentally, even the extraordinary amounts of money that were thrown at the Shuttle did not increase its safety; it is still the most dangerous rocket system ever created.

"What is the fastest way to become a commercial space millionaire? Start as a commercial space billionaire."

Hackneyed joke spawned by the number of companies that have tried and failed to launch rockets into Low Earth Orbit (LEO).

The few private companies that *had* managed to get something into orbit had used hardware developed under government programs, and their services were not cheap. To Musk, launch prices were a damning indictment of the state of space exploration, a business that had spent hundreds of billions of dollars on rocket technology with the result that the cost of putting something into LEO still cost around $10,000 per pound – before SpaceX came along. It was this lack of progress that particularly frustrated Musk, who decided he would aim to reduce those costs by half. Or more. To many space industry observers, it was a tall claim.

Musk knew the stakes were high. He knew little about the rocket industry and had never actually built anything – except Internet companies – in his life. The odds were hardly in his favor. But Musk had thrived in businesses where the default expectation was failure, so why not roll the dice on building rockets? The problem was how to do it. Musk started by going to the heart of the aerospace world in El Segundo, California, one of the beach cities just south of Los Angeles International Airport, where he began recruiting industry veterans for SpaceX.

One of his first hires was Tom Mueller, one of the world's leading propulsion experts. Designing propulsion systems had come naturally to Mueller, who came from a hands-on background. Mueller was born in St. Maries, Idaho, a logging community of about 2,500 people. His father was a log truck driver and he wanted his son to be a logger, so it was only natural that the younger Mueller grew up around logging trucks and chainsaws. It was an environment that spurred an interest in figuring out how things worked, which explains why he took his father's lawn mower apart. His dad was upset when found the parts, because he did not think he could put the pieces back together, but the younger Mueller re-assembled it and the machine ran pitch perfectly.

From rebuilding lawn mowers, Mueller moved on to building and flying toy rockets. He bought Estes rockets from his local hobby shop, although they did not last long because he usually crashed them or blew them up. In junior high, Mueller submitted a hybridized life sciences-propulsion project to the science fair, which was to fly an Estes rocket carrying crickets in it to see what the effects of acceleration were on the insects. Unfortunately, the parachute failed and the deceleration when the rocket hit the ground killed the crickets. Not wanting to kill anymore wildlife, Mueller restricted his next project to building a rocket engine out of his father's oxy-acetylene welder, making a rocket engine by injecting water into it to see what effect that had on its performance. The first time he ran it, the engine burned a hole through the side of the chamber, but with some minor modifications he was able to run it in a steady state, an achievement that allowed him to reach the regional round of the science fair.

Mueller earned a master's degree in Mechanical Engineering from the Frank R. Seaver College of Science and Engineering at Layola Marymount University, and received job offers for work in Idaho and Oregon, though they were unrelated to rocketry. So, Mueller decided to move to California to get a rocket job, eventually taking a position with TRW Space and Electronics where he spent 14 years running the Propulsion and Combustion Products Department. Along the way he earned the TRW Chairman's Award and filed several patents in propulsion technology. Mueller was happy working there, but his ideas about rocket engine design were lost in a company for which rocket engines were not a core component. To satisfy his creative impulses, Mueller turned to the Reaction Research Society, building his own engines and launching them in the Mojave Desert with fellow rocketeers. By 2002 he had almost completed the world's largest amateur liquid-fueled rocket engine, capable of producing 13,000 pounds of thrust. Musk met the enterprising propulsion engineer in January 2002, just as Mueller was preparing to attach his monster engine to an airframe. For Musk, building rocket engines was the key to his commercial spaceflight enterprise. He took one look at the rocket engine, and asked Mueller if he could build a bigger one.

Having recruited a slew of rocket engineers, Musk now needed someone to run day-to-day operations. Gwynne Shotwell (Figure 1.4) ran into Musk when she dropped off a friend, who had just started working at SpaceX, from lunch. The friend had mentioned to Musk that he should hire a business developer and Musk agreed, hiring Shotwell, who became the company's seventh employee, as Vice President of Business Development. In that position she built the Falcon vehicle family manifest to over 100 launches, representing over $3 billion in revenue. Today, as President, Shotwell is still the powerhouse of the company, responsible for day-to-day operations and for managing all customer and strategic relations to support company growth.

Figure 1.4. Gwynne Shotwell is President and Chief Operating Officer (COO) of SpaceX. Joining SpaceX in 2002 as Vice President (VP) of Business Development, Shotwell built the Falcon vehicle manifest to more than 100 launches. In 2018, Fortune Magazine listed Shotwell at #42 on their list of the World's 50 Greatest Leaders. Credit NASA.

Prior to joining SpaceX, Shotwell had spent more than ten years at the Aerospace Corporation, where she held positions in Space Systems Engineering and Technology and Project Management. She was promoted to the role of Chief Engineer of a medium launch vehicle-class satellite program, managed a landmark study for the Federal Aviation Administration (FAA) on commercial space transportation, and completed an extensive analysis of space policy for NASA's future investment in space transportation. Shotwell was subsequently recruited to be Director of Microcosm's Space Systems Division, where she served on the executive committee and directed corporate business development.

With his team in place, all Musk had to do was to get on with the business of building rockets… and rocket engines. Traditionally, rocket manufacturers bought engines from established companies, because the prospect of designing and building your own rocket engines was complex, time-consuming and expensive. But building his own rockets and engines was *exactly* what Musk intended to do. Not content with the challenges of establishing a rocket company, Musk also found time to co-found Tesla Motors[4] in 2004. One of the main objectives of Tesla Motors (named after electrical engineer and physicist Nikola Tesla) was to develop an environmentally-friendly sports car. To that end, the company built the snappy Tesla Roadster, a car that charges overnight, uses no gasoline, and sprints from zero to 96 kph in less than four seconds. This vehicle was followed by a family of all-electric cars, including the Model S (Figure 1.5) which became the company's flagship.

While Musk developed the Tesla venture, his SpaceX team began building a two-stage rocket, powered by 'Merlin', a compact, durable engine designed to lift a first stage, and 'Kestrel' for the second. The rocket, christened the Falcon 1 (Figure 1.6), was designed to lift 1,400 pounds into LEO. Getting the payload into orbit proved a tougher task than Musk had envisaged, however.

First, there was the issue of finding a launch site. Musk had originally hoped to launch the Falcon 1 booster (carrying a TacSat-1 satellite built by the U.S. Naval Research Laboratory) from Vandenberg Air Force Base (VAFB) in California, but that plan was stymied by a delay in launching a Titan-4 rocket carrying a classified payload. After spending an estimated $7 million on its VAFB facilities, SpaceX was told to leave the Complex 3 West launch site at Vandenberg. "It is just, I think, a travesty," Musk told *SPACE.com* in an interview at the 19th Annual Conference

---

[4]The Tesla enterprise started in 2003 when two teams, one consisting of Martin Eberhard and Marc Tarpenning, and the other of Ian Wright, JB Straubel and Musk, were trying to think of ways to commercialize the T-Zero prototype electric sports car created by AC Propulsion. One of Musk's goals had been to commercialize electric vehicles, starting with a premium sports car before moving to mainstream vehicles. It was suggested that the teams join forces to maximize the chances of success, so Musk became chair, Eberhard took on the role of Chief Executive Officer (CEO) and Straubel became Chief Technology Officer (CTO).

Figure 1.5. Tesla Model S. Credit. El Monty. Public domain.

Figure 1.6. Falcon 1 Flight 4 on September 29, 2008. Credit SpaceX.

on Small Satellites. Having signed an agreement with the U.S. Air Force to use Complex 3 West, and having made investments in the site as well as having paid for requisite environmental assessments, Musk had cause to be annoyed.

Fortunately, SpaceX had an alternate launch site on Omelek Island in the Kwajalein Atoll, a location that had been part of the company's plans to orbit payloads from there as well as from California and Florida. Kwajalein Atoll, which is part of the Republic of the Marshall Islands (RMI), lies in the Ralik Chain, 3,900 km southwest of Honolulu, Hawaii. On February 6, 1944, the atoll was claimed by the United States and was taken, together with the rest of the Marshall Islands, as a Trust Territory of the United States. In the years following the American invasion, the atoll was converted into a staging area for further campaigns in the advance on the Japanese homeland in the Pacific War, and as a command center for Operation Crossroads and for nuclear tests at the Marshalls atolls of Bikini and Eniwetok. Kwajalein Atoll is controlled by the United States military under a long-term lease and is part of the Ronald Reagan Ballistic Missile Defense Test Site.

Falcon 1 was shipped to Omelek Island by barge for a projected launch date of September 30, 2005, although Musk acknowledged delays were likely. While Omelek Island was remote, the location offered some advantages. To begin with, there are no population centers nearby, making range safety easier. Secondly, just about any orbit is achievable from Kwajalein, thanks to its proximity to the equator. Thirdly, any work done at Kwajalein only had to satisfy one entity – the Environmental Protection Agency – whereas in California, multiple federal agencies had to be engaged, along with state and county entities. The downside to launching from Omelek was the problem of hauling all the equipment needed to launch a rocket, as well as the challenge posed by the humidity, temperature and sea spray, which, in combination, created just about the most corrosive environment on the planet. It was this last factor that proved costly.

At about the same time Falcon 1 was being shipped to Omelek Island, and despite his company not having flown a single rocket, Musk had already signed three launch contracts (with the Swedish Space Corporation, MacDonald, Dettweiler and Associates Ltd – MDA – of Canada, and a commitment by an unspecified U.S. company), and had invested about $100 million in SpaceX. There was *a lot* riding on that first launch.

The 21-meter-high Falcon 1 was the first in a family of boosters planned by SpaceX to offer a more affordable option to launch satellites. Cost-capped at just $6.7 million, Falcon 1 launch vehicles were designed to carry up to 570 kilograms into LEO. Fueled by kerosene and liquid oxygen, the booster featured the SpaceX in-house-designed Merlin engine and a reusable first stage which, if everything went according to plan, would parachute back to the ocean for later recovery and reuse on a future flight. Falcon 1's first launch attempt came on

March 24, 2006 and Musk played down the chances of success, telling reporters that the likelihood of a new rocket launching from a new launch pad successfully on its first attempt was low. Musk's prediction proved right on the mark, because a fuel leak 25 seconds after launch caused a fire in the first stage, with the result that the 20-kilogram Falconsat fell through the roof of the SpaceX machine shop. "We had a successful liftoff and Falcon made it well clear of the launch pad, but unfortunately the vehicle was lost later in the first stage burn," Musk said in an update posted to the company's website. Shortly after the launch, and accompanied by engineer Tim Buzza, Mueller, the range safety officer, and the vice presidents of avionics and structures, Musk boarded a helicopter and flew over Omelek. Except for a fuel slick just offshore and a few scattered pieces of debris, there was little left of the rocket, the culmination of four years, tens of millions of dollars and endless seven-day work weeks. For the engineers, many of whom had quit steady jobs with Boeing and Lockheed, it was not the payoff they had been hoping for.

After poring over video footage, data points and flight telemetry, the cause of the launch failure was identified as a small fire that had broken out on the first-stage engine. The fire had been caused by a fuel leak, and the fuel leak was caused by the failure of an aluminum nut from the fuel pump. The nut had cracked, having been corroded in the salty, humid air. The choice to go with aluminum fittings rather than more durable stainless steel had been a cost-saving decision; in the business of launching rockets, weight equals money and aluminum is much lighter than steel. Unfortunately, that aluminum had been sitting in the humid tropical environment for ten weeks, with predictable – and costly – results.

Fixing the nut corrosion problem was simple; SpaceX replaced the aluminum with stainless steel. The team also added fireproof baffling around the engines and, as a further precaution against the heavy tropical air, kept the rocket inside a Quonset hut until a few days before liftoff. They also updated the launch software. Computers had recorded the fuel leak that destroyed the first Falcon 1 but nobody had noticed, so a new launch system was devised that would abort a countdown automatically in the event of an anomaly. In all, engineers made 112 changes to the rocket and the launch sequence.

Once the changes had been made, the modifications were tested rigorously and, nearly one year to the day after the failed first attempt, the rebuilt Falcon 1 was ready to go again. It stood on the launch pad under a blazing yellow sun, the first new launch system in 30 years. In many ways, Musk had already made history even before the launch; he had brought a privately built rocket to the launch pad twice in one year, something no one else had ever done. Launch attempt #2 took place on March 21, 2007, but liquid oxygen slosh occurred in the second stage which triggered an oscillation 90 seconds into the burn. The instability continued for another 30 seconds and caused the stage to enter an uncontrollable roll that exceeded the design limits of the second stage roll control thrusters. The roll torque caused propellant to be forced away from the engine intakes, with the result that the Kestrel engines, now starved of fuel, simply flamed out. A note was made by the SpaceX engineers to include second stage slosh baffles in future flights.

After replacing the first stage, the scene was set for Falcon 1 to be launched on August 3, 2008. Despite the first two scrubs, Musk was unfazed, telling reporters that hiccups were to be expected with the debut of any new launch system and adding that each launch attempt brought valuable experience to the flight team. True enough, but it was experience that the flight team was growing weary of. This time, the new Merlin 1C engine completed its burn, but the thrust took longer to decay than planned. This meant that the engine was still providing thrust when the second stage tried to separate, with the inevitable result that the two stages collided. The second stage was sent tumbling out of control.

September 28, 2008, was the date set for launch attempt #4. This time, the launch proceeded according to plan. One minute before launch, the Falcon 1 switched to the computerized launch sequence. Seconds before launch, a spark lit and fired a turbo pump spinning at 21,000 rpm, pushing LOX and kerosene into the Falcon's main engine combustion chamber. The Falcon launch vehicle was on its way. During the flight, the second stage demonstrated restart capability and the payload was delivered into its intended orbit.

By this point, Musk was breathing rarefied air. He was living in a 6,000-square-foot house with a domestic staff of five in Bel Air hills, attending black tie fundraisers, partying with Leonardo DiCaprio and Paris Hilton, and jetting to Richard Branson's Necker Island on a private jet. He also served as the inspiration for the Tony Stark character of *Iron Man* fame. When development of the first *Iron Man* movie was in its infancy, director Jon Favreau had a problem because he could not seem to bring his main character, egotistical genius/superhero Tony Stark, to life. In search of inspiration, Favreau turned to Robert Downey Jr., the actor hired to play Stark/Iron Man. He recommended that the two should sit down with Musk and the rest, as they say, is history; the comparison was spot-on. For those who are film buffs, the first *Iron Man* film briefly featured a Tesla Roadster parked in Stark's underground shop, while in the sequel Musk made a cameo appearance with Stark, asking if he might be able to design an electric jet. Musk was living a dream lifestyle, but it was also a very busy one. Aside from serving as a muse for an iconic comic book hero, Musk was also overseeing the developmental progress of the Falcon 1, breaking ground at Launch Complex 40 at Cape Canaveral Air Force Station, monitoring the progress of the company's multi-engine test-firing, and of course, running Tesla.

Tesla hit a small hurdle in January 2008, when the company fired several key personnel to pare down costs after a performance review by Ze'ev Drori, Tesla's new CEO. The following month, a fifth round of funding added another US$40 million (Musk had contributed US$70 million of his own money by this time). Another hurdle of the personal variety struck Musk in September 2008, when his wife, Justine, announced she was divorcing him. They entered counseling, but with Tesla and SpaceX to run Musk had precious little time to solve marital issues and, after three sessions, he filed for divorce.

While dealing with the divorce, Musk had his attention focused on an impending launch attempt of the Falcon 1, which was being prepared at the U.S. Army Kwajalein Atoll site. But the media was not following the Falcon 1 story; thanks to Musk's wife being a prolific blogger, the public was treated to a blow-by-blow account of Musk's divorce proceedings. Predictably, the media followed the story and embellished and propagated it in a way only the news media knows how. Some newspapers suggested the reason for Musk's divorce was his relationship with actress Talulah Riley, but Musk had filed for divorce before he had met Riley. Other, equally creative news entities suggested that Matt Peterson, the president of Global Green, and Justine's long-term boyfriend, had played a part in Musk's decision to file for divorce. Again, the reality was markedly different; Peterson had been a mutual acquaintance of Musk and his wife for many years, and Musk's wife did not enter a relationship with him until after Musk had filed for divorce. The bottom line in the divorce was that there was no third party involved in the break-up at all. Still, that did not stop CNBC from featuring the Musk divorce on *Divorce Wars*, a reality TV show that dissected the break-up piece by piece, explaining to those who were interested in messy divorces that Musk's wife was looking for $6 million in cash and stock.

After Musk's divorce, and the successful launch of the Falcon 1, a filing from the settlement was circulated which stated that Musk was broke and living on loans. The news was ammunition for Musk's critics, who suggested that since Musk was unable to manage his own money, he was probably unsuited to running a multi-billion-dollar car company. As with many venture capitalists, Musk's $1.9 billion net worth was mostly tied up in his ventures, including SpaceX. Like the spurious divorce stories, the 'broke millionaire' story was equally erroneous. Musk had had to take a tough decision in late 2007 when Tesla was in a serious financial state, and the only way out was for existing shareholders to recapitalize the company. Because the company was in such dire straits there was no way to raise money externally and, rather than allow Tesla to die, Musk committed almost all of his cash reserves to the company, leaving a few million dollars to cover living expenses. Of course, this reality did not satisfy the media, who targeted Musk's private jet as an extravagant indulgence and portrayed Musk as a playboy millionaire. Once again, the media accounts were way off the mark. To begin with, Musk did not own any homes or expensive yachts, and hardly ever took vacations. He still does not. He is a self-confessed workaholic. As for the jet, in an average year

Musk makes more than 200 business trips, spending 500+ hours in the air. That crushing schedule would be difficult to maintain without a jet.

By January 2009, Tesla, which had raised US$187 million and delivered 147 cars, had attracted the interest of Germany's Daimler AG. On May 19, 2009, the manufacturer of Mercedes acquired an equity stake of less than ten percent of Tesla for a reported US$50 million. In July 2009, shortly after Tesla was approved to receive US$465 million in interest-bearing loans from the United States Department of Energy (DoE), the company announced it had achieved corporate profitability for that month, having earned approximately US$1 million on revenue of US$20 million. More good news came that same month, following the second successful launch of a Falcon 1 carrying the RazakSAT satellite. Later that year, a new investor – Fjord Capital Partners, a specialized European private equity manager investing in the clean energy sector globally – came on board, and early in 2010 Tesla Motors indicated its intention to file an Initial Public Offering (IPO).

Shortly before its IPO, Musk hit the headlines again. This time the subject was not his divorce or SpaceX, but corporate excess. In a U.S. Securities and Exchange filing, Tesla disclosed it had paid $175,000 in fuel charges and landing fees associated with Musk's private jet. Now, $175,000 is hardly an astronomical sum, but at a time when the public was particularly sensitive to corporate excess (remember, this was when certain motor companies were receiving government bailouts), the news media asked why Musk could not have picked up the tab himself. There was no suggestion that Musk had run afoul of any laws because all the flights were for company business, and it was hardly unusual for CEOs to fly on private jets funded by their companies. The $175,000 that Musk had billed the loss-making company had included trips to Washington, D.C., to secure the project's $465 million DoE loan. Some may remember the rebuke from Congress that was provoked when the Big Three Detroit CEOs flew to Washington by private jet to ask for government help. As it geared up for an IPO, Tesla should have known that it would be subject to increased scrutiny and, given the motor industry controversy, perhaps Musk had failed to anticipate how such flights might be perceived.

With the successful launch of Falcon 1, Musk could look to the future, a large part of which was figuring out a way to land humans on Mars. To achieve that goal, Musk knew he first had to establish a blueprint that followed a path of ever-increasing capability and reliability while simultaneously reducing costs. By following this plan, Musk ultimately hoped to reduce the cost of launching vehicles into orbit and increase the reliability of space access by a factor of ten. It was a bold ambition, but Musk believed that by eliminating the traditional layers of internal management and external sub-contractors, SpaceX could reduce its costs and accelerate decision-making and delivery. He also believed that by keeping most of the manufacturing in-house, SpaceX could further reduce its costs, keep tighter control of quality, and ensure a tight feedback loop between the design and manufacturing teams. By focusing on simple, proven designs with a primary emphasis on reliability, Musk was sure SpaceX could bring down the costs associated with complex systems operating at the margins. He needed to do this to make money to fund his crewed Mars mission. Coupled with the newly emerging market for private and commercial space transport, Musk also hoped that his new model for reaching orbit would re-ignite humanity's efforts to explore and develop space, and Mars in particular.

But in 2018, two years after announcing the Big Falcon Rocket (BFR), Musk was once again struggling with the fallout of Tesla. Production targets for the Model 3 had been pushed back several times and customers, who had placed $1,000 deposits in 2016, were growing impatient. Then, in June 2018, Tesla announced it was laying off nine percent of its 3,000+ workforce. While the 2018 situation Musk faced with Tesla was less serious than the one he had faced ten years earlier, the risk was more pronounced because Tesla and SpaceX were established companies. Any failure of Tesla and/or SpaceX would risk thousands of jobs and billions of dollars. In addition, during those intervening years, Musk's space company had not only become an integral element of the US space program but also a critical means for countries and companies around the world to launch satellites.

As the chief executive of Tesla and SpaceX, and as the proponent behind the Hyperloop transit system, Neuralink, OpenAI and crewed missions to Mars, Musk is idolized in Silicon Valley, but there are others who wonder if the Mars Messiah takes on too much. This concern is not just borne of Musk's willingness to take on grandiose projects, but from his habit of setting outrageously ambitious deadlines that he often fails to meet; in the summer of 2018, his Tesla Model 3 had a waiting list that stretched to almost half a million. Then there is that target of humans on Mars by the mid 2020s. Is it doable? Musk thinks so, but at what risk?

"For the early people that go to Mars, it will be far more dangerous. It kind of reads like Shackleton's ad for Antarctic explorers: difficult, dangerous, good chance you will die. Excitement for those who survive."

Quote from an interview with Elon Musk by Thomas Barrabi, Elka Worner, March 13, 2018.

The vehicle for that trip was announced in September 2016. Dubbed the Interplanetary Transport System (ITS), later rebranded the Big Falcon Rocket (BFR), and eventually morphing into the Starship (Figure 1.7), this monstrous launch vehicle is the personification of one of the world's brashest and most ambitious entrepreneurs. It is also the embodiment of the *raison d'être* for SpaceX, the company whose story is told in the chapters that follow.

Figure 1.7. The SpaceX Human Lander System (HLS) Starship is designed to land on the Moon, but the vehicle's architecture is intended to evolve into a fully reusable launch and landing system designed for travel to Mars. Credit NASA.

2

The Engine of Competition

"With the advent of the ISS, there will exist for the first time a strong, identifiable market for 'routine' transportation service to and from LEO, and that this will be only the first step in what will be a huge opportunity for truly commercial space enterprise. We believe that when we engage the engine of competition, these services will be provided in a more cost-effective fashion than when the government has to do it."

Dr. Michael D. Griffin, November 2005

Figure 2.0. May 25, 2012. A SpaceX Dragon commercial cargo vehicle is grappled by the Canadarm2 robotic arm at the ISS. Expedition 31 crewmembers Don Pettit and Andre Kuipers grappled Dragon at 9:56 a.m. (EDT) and used the Canadarm2 to berth Dragon to the orbiting outpost's Harmony Module. Dragon became the first commercially developed space vehicle to be launched to the ISS. Credit NASA.

© Springer Nature Switzerland AG 2022
E. Seedhouse, *SpaceX*, Springer Praxis Books,
https://doi.org/10.1007/978-3-030-99181-4_2

Following the spectacular success of the SpaceX Dragon flight to the International Space Station (ISS) in May 2012 (Figure 2.0), it was of little surprise that the Obama Administration tried to take some of the credit. When Presidential Science Advisor John Holdren declared that the Obama Administration had made the Dragon flight possible, there were some rumblings among Republicans, who felt justifiably aggrieved since the Commercial Orbital Transportation Services (COTS) program had been proposed by the Bush Administration in 2005, and the COTS contract that funded the SpaceX mission had been awarded in 2006. On the issue of giving credit where credit is due, Michael Griffin, who led NASA during the Bush Administration, should also be mentioned, since it was Griffin who conceived and funded the COTS federal seed money program that finally got Dragon off the ground. However, the Obama Administration should still receive some recognition because it was President Barak Obama who increased commercial funding to $500 million a year.

> "*I would like to start off by saying what a tremendous honor it has been to work with NASA. And to acknowledge the fact that we could not have started SpaceX, nor could we have reached this point without the help of NASA.*"

Elon Musk, at a press conference after the launch of Dragon's first ISS flight.

You may wonder why Musk thanked the space agency so profusely following the Dragon's historic flight to the ISS. After all, there were some who believed that Musk was working to privatize NASA out of existence. You may also wonder how much help NASA provided to SpaceX, since the public perception, even ten years after Dragon's docking with the ISS, is still that the commercial space race is being funded by the companies themselves. Dragon's mission was not a libertarian adventure. Far from it. It was, in fact, the result of a collaborative effort between SpaceX and NASA. So, the purpose of this chapter is to provide an insight into the various funding initiatives that have helped SpaceX and the other companies in the commercial space arena, and to make sense of all those myriad funding acronyms like CCDev, COTS, CRS, and SAA.

We will start with those Space Act Agreements (SAA). In 2003, when it was announced that the Space Shuttle (Figure 2.1) was to be retired, NASA had to figure out how to continue to transport cargo and humans to the ISS. Fortunately, the technology needed to launch to Low Earth Orbit (LEO) was firmly established, and the private sector seemed to be ready to step up to the task. So, in 2006, NASA began investing in private spaceflight companies through a program known as COTS. The purpose of COTS was to coordinate the development of vehicles to ferry cargo and crew to the ISS. At the time of the program's announcement on January 18, 2006, NASA anticipated that commercial services to the orbiting outpost would be required until 2015. Driving the COTS program were the SAAs. The SAAs were driven by milestone-based payments and the milestones were provided by NASA. The difference between the SAAs and the

Figure 2.1. Space Shuttle *Discovery* and her crew begin the STS-121 mission. Credit NASA.

Commercial Resupply Services (CRS) program, which we will get to shortly, is that COTS did not involve binding contracts, whereas the CRS program did. As for the difference between COTS and CRS, one (COTS) was all about *vehicle development* and the other (CRS) was all about *deliveries*. A third program, Commercial Crew Development (CCDev) was a related program regarding *developing crew services*. All three programs are managed by NASA's Commercial Crew and Cargo Program Office (C3PO).

This was not the first time that government had given industry a helping hand; after all, it was the U.S. government's early support of the railroad and the aviation industry that laid the foundation for private companies to succeed. Nevertheless, for the space industry, the agency's agreements with its newer partners such as SpaceX represented a sea change in the way it worked with the private sector. Rather than the traditional cost-plus model, in which companies were reimbursed the *cost* of a project *plus* an additional amount that guaranteed a profit, SpaceX and its competitors worked under SAAs in which NASA paid increments of a fixed price once the companies accomplished agreed upon milestones (Figure 2.2). In short, the companies only got paid for what they achieved. If they did not hit the milestones, they did not get paid.

Milestone 1: Project Management Plan Review Subsequent to Space Act Agreement execution and initiation of the COTS program, SpaceX shall host a kickoff meeting to describe the plan for program implementation, which includes management planning for Design, Development, Testing, & Evaluation (DDT&E), integrated schedule, financing, supplier engagement, risks and anticipated mitigations. SpaceX shall provide a briefing of the program implementation plan, along with a hard copy of the presentation materials, and responses to any questions that the NASA Team might have concerning SpaceX's plan. Acceptance within 5 days and payment within 15 days Success Criteria: Successful completion of the project management plan review as described above.	Amount: $23,133,333 Date: Sept. 2006
Milestone 2: Demo 1 System Requirements Review SpaceX shall conduct a Demonstration 1 System Requirements Review in accordance with the SRR definition in Appendix 3. Success Criteria: Successful completion of the SRR.	Amount $5,000,000 Date: Nov. 2006
Milestone 3: Demo 1 Preliminary Design Review (PDR) SpaceX shall conduct a PDR in accordance with the PDR definition in Appendix 3 Success Criteria: Successful completion of the PDR.	Amount: $18,133,333 Date: February 2007

Figure 2.2. Examples of milestones and success criteria for the Cargo Dragon. Credit NASA

Another distinctive element of the SAAs is that they left spacecraft design mostly in the hands of private companies. So, once the SAA was awarded, NASA did not tell SpaceX (or anyone else) whether it was doing something right or wrong; it just let them get on with the business of designing and building spacecraft. There was nothing stopping SpaceX from asking NASA for advice (it often did) but the arrangement meant that SpaceX did not build to any pre-established design specifications. For the most part, SpaceX was left to its own devices to figure out how best to achieve the agreed upon milestones. The SAA system is effective because the approach encourages innovation, reduces NASA's renowned tendency to micromanage and, perhaps most importantly, saves money. In fact, NASA reckoned SpaceX was able to build its Falcon 9 rocket for one-third of what the agency would have spent on a similar project under its traditional model.

The SAAs have been a boon for SpaceX and the commercial spaceflight industry, since NASA has used these agreements to funnel millions of dollars of investment into the companies. The agreements have been around for a while (NASA was first given the authority to enter into such agreements under the Space Act of 1958) and there are several ways they can be used. Before the SAAs, the traditional government method of awarding a contract involved a request for bids for a specific project, with cost overruns being paid by the government or the contractor according to a legal document. This method rarely worked well and, in the 'learn-as-you-go' spaceflight arena, such a system has often been detrimental to the design and development of space systems and associated hardware. So the SAAs were created to harness collaboration between government, industry, and academic researchers. Today there are three basic types of SAA: reimbursable, non-reimbursable, and funded agreements. Each of these are described here.

Under *reimbursable* agreements, a commercial firm or academic researchers may access 'unique goods, services or facilities' that NASA possesses but is not fully using. In this type of SAA, a company might reimburse NASA for service use, because in return it does not have to invest in those services alone. The *non-reimbursable* SAA, as its name suggests, is one in which no money is reimbursed. For example, in one instance a university wanted to conduct tests on the effects of collisions of cometary matter. Under a non-reimbursable SAA, the university worked with Johnson Space Center (JSC) free of charge because NASA also needed a better way to analyze data from its Stardust and Deep Impact comet missions. For *funded* SAAs, NASA only pays if a company reaches each of several agreed upon milestones (we will discuss these later) in design, safety, and performance, by a specific date. For example, before Dragon flew to the ISS, SpaceX had to complete a Critical Design Review (CDR, Figure 2.3) of the vehicle. By completing the review successfully, SpaceX received a $25 million payday.

Since retiring the Shuttle in 2011, NASA has increasingly used funded SAAs to spur the development of manned and unmanned spacecraft, to realize its goal of finding a homegrown way to send crew and cargo to the ISS. In the longer-term, NASA hopes the investment in innovation and infrastructure will foster a competitive market that can keep costs down for future missions in LEO and deep space. The bottom line is that these SAAs help NASA keep costs down if project delays cause budget overruns, because the extra expenses are borne by the firm, not the taxpayer. This is good reasoning for those who must explain government costs to wallet-weary taxpayers, and it is also good for NASA because the agency can defend the use of the agreements to stimulate innovation and growth in the commercial spaceflight industry.

Critical Design Review	
Entrance Criteria	Success Criteria

1. Successful completion of the PDR and responses has been made to all PDR open issues, or a timely closure plan exists for those remaining open.
2. A preliminary CDR agenda, success criteria, and charge to the board have been agreed to by the technical team, project manager and review chair prior to the CDR.
3. CDR technical products listed below for both hardware and software system elements have been made available to the cognizant participants prior to the review:
  a. Updated baselined documents, as required.
  b. Product build-to specifications for each hardware and software configuration item, along with supporting trade-off analyses and data.
  c. Fabrication, assembly, integration, and top-level test plans and procedures.
  d. Technical data (e.g., Integrated Schematics, Spares Provisioning List, engineering analyses, specifications, etc.).
  e. Interface Control Documents (e.g. Command and Telemetry List, instrumentation, electrical, mechanical, fluids & gas interfaces, user interfaces)
  f. Preliminary Test Requirements document (e.g. Operational Limits and Constraints, acceptance criteria)
  g. Verification & Validation Plan (including requirements and specification).
  h. Launch Site Operations Plan, including Checkout and Activation Plan.
  i. Updated risk assessment and mitigation.
  j. Updated schedule data.
  k. Updated logistics documentation.
  l. Software Design Review
  m. Updated LLIL.
  n. Subsystem-level and preliminary operations hazards analyses.
  o. Systems and subsystem certification plans and requirements (as needed).
  p. System hazard analysis with associated verifications.
	1. The detailed design is expected to meet the requirements with adequate margins at an acceptable level of risk.
	2. Interface control documents are appropriately matured to proceed with fabrication, assembly, integration and test, and plans are in place to manage any open items.
	3. High confidence exists in the product baseline, and adequate documentation exists and/or will exist in a timely manner to allow proceeding with fabrication, assembly, integration, and test.
	4. The product verification and product validation requirements and plans are complete.
	5. The testing approach is comprehensive, and the planning for system assembly, integration, test, and launch site and mission operations is sufficient to progress into the next phase.
	6. Adequate technical and programmatic margins and resources exist to complete the development within budget, schedule, and risk constraints.
	7. Risks to mission success are understood, and plans and resources exist to effectively manage them.
	8. Safety and mission assurance (i.e., safety, reliability, maintainability, quality, and EEE parts) have been adequately addressed in system and operational designs and any applicable S&MA products (i.e., hazard analysis and failure modes and effects analysis) have been approved.

Figure 2.3. Critical Design Review criteria for Dragon. Credit NASA.

Thanks to the SAAs, SpaceX has been able to leapfrog ahead of where they would have been had they tried to fly their Dragons (Cargo Dragon and Crew Dragon) to the ISS in a purely commercial environment. But money is not the only reason SpaceX and other private spaceflight companies still clamor for government support. Agreeing on an SAA partnership with NASA confers a great source of legitimacy for an industry that, for some people, still smacks of science fiction. That legitimacy is vital for attracting other customers, as well as more traditional investors from Wall Street and investment banks. An added bonus is that NASA also helps private spaceflight companies navigate the complicated regulatory environment, assistance which was especially important when the safety of its astronauts became an issue when certifying Crew Dragon and the Boeing Starliner.

The SAAs also benefit the agency because LEO is no longer a frontier after 60 years of successful missions, and NASA has set its sights on destinations further afield such as Mars. But to get to Mars means developing a super-heavy-lift launch vehicle (the Space Launch System – SLS) *and* a new interplanetary spacecraft (the

Orion). The development of the SLS and Orion has cost an obscene amount of money. Orion, the prime contractor for which is Lockheed Martin, has been in development since 2006. The spacecraft is an orphan of the Constellation Program conceived to ferry astronauts to the ISS and ultimately to the Moon and was planned to be launched atop an Ares rocket. Constellation was cancelled by the Obama Administration, but Congress decided to keep Orion (also known as the Multi-Purpose Crew Vehicle – MPCV) and replace Ares with the SLS. Including the $6.3 billion spent on Orion's development during the Constellation Program, Orion (Figure 2.4) has cost $12.2 billion to date. NASA's Office of the Inspector General (OIG) estimates Orion's Life Cycle Cost through 2030 to be $29.5 billion, a figure that does not include the costs of the service modules. As mentioned, an *obscene* amount of money. Yet despite this already substantial expenditure the Orion program continues to slip, with a first manned flight not expected until August 2023 at the earliest. That is 17 years to develop a spacecraft. Little wonder that there are so many fans of SpaceX, which was awarded just $1.75 billion to develop Crew Dragon.

Figure 2.4. Orion. The $29.5 billion spacecraft: the next giant leap in excessive spending, with no end in sight. Credit NASA.

Clearly, NASA simply cannot afford to develop these vehicles *and* maintain the ISS *and* support crew and cargo missions to the station, so the plan all along was to hand over transportation to and from the ISS to private companies. By doing this, the agency has been financially freed to develop the SLS (Figure 2.5) that will be used to travel to the Moon and ultimately to Mars.

Figure 2.5. The SLS is a super-heavy-lift launch vehicle that will send astronauts to the Moon and eventually to Mars.

"For 50 years, American industry has helped NASA push boundaries, enabling us to live, work and learn in the unique environment of microgravity and low Earth orbit. We're counting on the creativity of industry to provide the next generation of transportation to low Earth orbit and expand human presence, making space accessible and open for business."

NASA's William Gerstenmaier.

Having outlined SAAs and examined the exorbitant costs of developing spacecraft, we will now take a closer look at COTS (Figure 2.6). When the agency was forced to retire the Shuttle during the Bush Administration, NASA suggested that commercial cargo services to ISS would be necessary through at least 2015. But even with the successes of SpaceX and the development of Crew Dragon and

Boeing's CST-100, that date slipped. Fortunately, thanks to various NASA funding programs such as COTS, at least the agency now has a choice of systems to ferry cargo and astronauts to the space station.

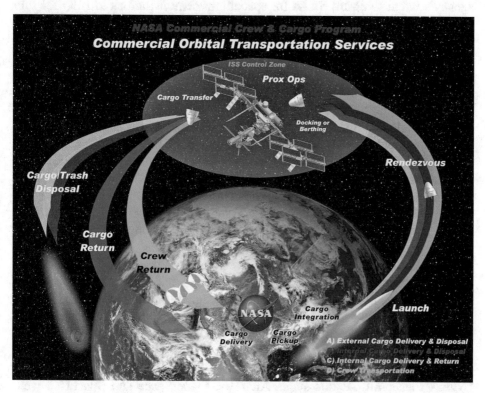

Figure 2.6. In 2004, President George W. Bush established the U.S. Space Exploration Policy (SEP). The SEP outlined a plan for returning to the Moon by 2020 and a deadline for the Shuttle to be retired by the end of 2010 after assembly of the ISS had been completed. A year later, Mike Griffin was appointed NASA Administrator, and one of his first tasks was to challenge U.S. private industry to develop ISS cargo and crew transportation capabilities. To give the commercial sector a helping hand, Griffin allocated $500 million over five years to stimulate the development and demonstration of commercial capabilities.

COTS is not the first time NASA has explored programs for ISS services. In the mid 1990s, NASA funded Alternate Access, a preliminary study which convinced many entrepreneurs that the ISS would emerge as a significant market opportunity. While the ISS market never transpired, the Shuttle's retirement forced NASA to revisit the purchase of commercial orbital transportation services on foreign spacecraft because there were no homegrown replacements in the pipeline. These foreign vehicles included the Russian Federal Space Agency's venerable Soyuz and Progress spacecraft, the European Space Agency's (ESA) Automated Transfer Vehicle (ATV), and the Japan Aerospace

Exploration Agency's (JAXA) H-II Transfer Vehicle (HTV). To bridge the gap to Commercial Crew, NASA initiated COTS, the first round of which took place in early 2006. In this round, NASA released an announcement that described the agency's intent to award SAAs for spaceflight demonstrations to LEO for four capabilities: external cargo delivery and disposal; internal cargo delivery and disposal; internal cargo delivery and return; and crew transportation. There were 21 proposals, submitted by 20 companies, of which 18 met the requirements for initial screening. By May 2006, six finalists had been selected, of which SpaceX was identified as being the clear leader not only for its technical strengths but also thanks to the company's finances.

Having selected SpaceX as its primary COTS partner, NASA had to decide which company would be the second partner. The choice was between Rocketplane Kistler and SpaceDev, with the former ultimately winning the decision and $207 million of funding (SpaceX was awarded $278 million). In 2007, Rocketplane Kistler had its SAA terminated with the company having failed to meet agreed upon technical and financial milestones. Shortly thereafter, COTS Round 2 was announced. This round allowed previous competitors to re-submit updated proposals and new competitors to propose plans. In this round, NASA received 13 proposals, four of which were eliminated due to business plan deficiencies. SpaceX, which also submitted a plan, had its proposal dismissed because it was one that NASA could have executed itself using its own funding. The eight proposals still on the table were evaluated using the same criteria as Round 1. Five of these were selected to move on to the next step: Andrews Space, Boeing, Orbital Sciences Corporation (Orbital), PlanetSpace and SpaceHab. A second stage selection in early 2008 was made to Orbital for its Cygnus spacecraft. At this stage, Andrews, PlanetSpace and SpaceHab were eliminated due to financial issues.

The COTS program concluded at the end of 2013 after SpaceX and Orbital had designed, built and launched spacecraft on launch vehicles that were new designs. NASA lauded the program as a success, but to understand just how successful the program was requires an insight into the machinations of government procurement. When NASA initiates a procurement, it must abide by the Federal Acquisitions Regulations (FARs). The FARs are codified at Title 48 of the Code of Federal Regulations (CFRs), which specify the myriad types of contracts and clauses that must be written into a procurement contract. For a launch capability, which is an extremely complex procurement, NASA pays the contractor all development costs and owns the resulting design. COTS, however, was not a procurement but was more of a partnership or transaction that was enabled by using the SAAs. To keep within the legal boundaries of oversight provided by the FARs, NASA had to employ some legalese, using the phrase 'Announcement for Proposals' instead of the usual 'Request for Proposal' when asking for bids. In addition to the legalities, COTS differed from traditional procurements because the companies selected had to pay part of the development cost. Another difference was the use of those milestone-based payments, a tool that helped reduce oversight because NASA did not have to specify the requirements of the launch vehicles. This allowed companies to move at their own speed, which was perfect for a nimble start-up like SpaceX.

For an idea of just how successful COTS was, bear in the mind the following numbers: the total funding for the program was $800 million, and for that money NASA gained access to two launch vehicles for ferrying cargo to the ISS. The SLS, albeit a more powerful launch vehicle, has been procured using conventional means and it has hoovered up more than $20 *billion* to date. It still is yet to fly. Another example is the Space Shuttle program, which cost $221 billion for just 135 flights. The Space Shuttle was a program that promised safe and reliable access to space, but ultimately 2.7 percent of the vehicles were lost in accidents. However, spaceflight numbers pale in comparison with military procurement. The F-35 was estimated to cost $200 billion in total acquisition cost, but this number has doubled and does not even include the trillion plus dollars required for long-term operations. Perhaps the greatest boondoggle of all is the Ford aircraft carrier, which has been described as nearly useless by the admirals charged with deploying it.

Without the COTS program, it is unlikely the Falcon 9 would have served as a catalyst for the commercial space revolution, and it is unlikely we would have been treated to the sight of reusable stages landing on drone ships. COTS was so successful that NASA decided to use the same procurement process for the Commercial Crew Program, which ultimately led to the successful flight of Crew Dragon to the ISS.

> *"Few would have imagined back in 2010 when President Barack Obama pledged that NASA would work 'with a growing array of private companies competing to make getting to space easier and more affordable,' that less than six years later we'd be able to say commercial carriers have transported 35,000 pounds of space cargo (and counting!) to the International Space Station – or that we'd be so firmly on track to return launches of American astronauts to the ISS from American soil on American commercial carriers. But that is exactly what is happening. Today's announcement is a big deal that will move the president's vision further into the future."*

Charles Bolden, NASA Administrator

Whereas the COTS program was all about vehicles and did not involve binding contracts, the CRS program (Appendix II) was all about deliveries and *did* involve legally binding contracts. The development of the CRS program also began in

2006, with the purpose of creating American commercially operated uncrewed cargo vehicles to service the ISS. Development of these cargo-carrying vehicles was under a fixed price milestone-based program, which meant each company receiving funding had to meet a list of milestones with a dollar value attached to them and those companies only received the funding if they achieved the milestones. The first of these CRS contracts was issued on December 23, 2008, when NASA awarded contracts to SpaceX and Orbital Sciences Corporation – Orbital. Under these contracts, SpaceX was to use its Falcon 9 rocket and Dragon spacecraft to haul cargo to the ISS, while Orbital would use its Antares rocket and Cygnus spacecraft to do the same. The contracts, worth a combined $3.5 billion through 2016 ($1.6 billion to SpaceX and $1.9 billion to Orbital), were awarded based on the likelihood of rocket availability and the superior management structures and technical abilities demonstrated by the two companies' proposals. In CRS phase 1, which was planned to run through 2016, SpaceX was to fly 12 cargo Dragon flights and Orbital was to fly eight Cygnus flights. The first of these resupply missions was flown by SpaceX in 2012 (Figure 2.7a, b, c). Three years later NASA extended Phase 1 by adding three flights to the contracts of both SpaceX and Orbital. A further extension later added five more flights to the SpaceX Phase 1 contract, for a total of 20 flights (see Table 2.1).

Figure 2.7a.   CRS-1 Mission Patch. Credit NASA

Figure 2.7b. CRS-1 Launch. October 8, 2021. Credit NASA

Figure 2.7c. The Dragon spacecraft makes its approach to the ISS in preparation for being grappled by the Canadarm2 operated by Don Pettitt. Credit NASA.

Table 2.1: List of SpaceX CRS Phase 1 Cargo Resupply Flights

Flight	Date
SpaceX CRS-1	October 8, 2012
SpaceX CRS-2	March 1, 2013
SpaceX CRS-3	April 18, 2014
SpaceX CRS-4	September 21, 2014. [1]
SpaceX CRS-5	January 10, 2015
SpaceX CRS-6	April 14, 2015
SpaceX CRS-7	June 28, 2015. [2]
SpaceX CRS-8	April 8, 2016
SpaceX CRS-9	July 18, 2016
SpaceX CRS-10	February 19, 2017
SpaceX CRS-11	June 3, 2017. [3]
SpaceX CRS-12	August 14, 2017
SpaceX CRS-13	December 15, 2017
SpaceX CRS-14	April 2, 2018
SpaceX CRS-15	June 29, 2018
SpaceX CRS-16	December 5, 2018
SpaceX CRS-17	May 4, 2019
SpaceX CRS-18	July 25, 2019
SpaceX CRS-19	December 5, 2019
SpaceX CRS-20	March 7, 2020

Notes:
1. Capsule reused.
2. Launch failure: 139 seconds into the launch, a rapid loss of pressure occurred in the liquid oxygen (LOX) tank of the second stage. The Falcon 9 broke up a few seconds later and the capsule impacted the ocean. The cause of the accident was traced to the failure of a strut that secured pressurized helium inside the LOX tank.
3. CRS-4 capsule was re-flown.

CRS-2 began with Orbital, SpaceX, Sierra Nevada Corporation (SNC), Boeing and Lockheed Martin submitting proposals. Three companies were awarded contracts in January 2016 – SNC for its Dream Chaser spacecraft, SpaceX for its Dragon 2 spacecraft and Orbital for its Cygnus. The maximum contract value was up to $14 billion and the contracts included several requirements, some of which are listed in Table 2.2.

Table 2.2: CRS Phase 2 Contract Requirements[1].

Deliver 14,000 to 17,000 kg per year, 55 to 70 m³ of pressurized cargo.
Deliver 24–30 powered lockers per year.
Deliver 1,500 to 4,000 kg per year of unpressurized cargo
Return/dispose of 14,000 to 17,000 kg per year, 55 to 70 m³ of pressurized cargo
Dispose of 1,500 to 4,000 kg per year of unpressurized cargo, consisting of 3 to 8 items

1. Source: International Space Station Commercial Resupply Services 2 Industry Day. ppt file. NASA. April 10, 2014. Archived from the original on April 3, 2015. Retrieved April 12, 2014.

Each of the companies was guaranteed at least six cargo missions under the CRS Phase 2 contract. For SpaceX, the CRS Phase 2 contract flights began in December 2020 (see Table 2.3).

Table 2.3: SpaceX CRS Phase 2 contract flights

Flight	Date
SpaceX CRS-21	December 6, 2020
SpaceX CRS-22	June 3, 2021
SpaceX CRS-23	August 29, 2021
SpaceX CRS-24	December 21, 2021
SpaceX CRS-25	May 2022 (planned)
SpaceX CRS-26	September 2022 (planned)
SpaceX CRS-27	January 2023 (planned)
SpaceX CRS-28	June 2023 (planned)
SpaceX CRS-29	October 2023 (planned)

The third phase of NASA's commercial development of space was Commercial Crew Development, or CCDev (see Appendix III). The intent of this multiphase technology development program, which is administered by the agency's C3PO, was to stimulate the development of privately-operated crew vehicles to LEO. Under the program, at least two providers would be chosen to deliver crew to the ISS. Ultimately, those two providers were SpaceX and Boeing. Unlike the traditional space industry contractor funding used on the Space Shuttle, Apollo, Gemini, and Mercury programs, funding for the CCDev program contracts was explicitly designed to facilitate only specific subsystem technology development objectives that NASA wanted for NASA purposes; all other system technology development was funded by the commercial contractor.

In the program's first phase (CCDev 1), NASA provided $50 million during 2010 to five companies, intended to foster research and development into human spaceflight concepts and technologies in the private sector. Later that year, a second set of CCDev proposals were solicited by NASA for technology development project durations of up to 14 months. The proposals selected included Blue Origin, which was awarded $3.7 million to develop an innovative 'pusher' Launch Abort System (LAS) and composite pressure vessels. Boeing received $18 million for the development of its Crew Space Transportation (CST)-100 vehicle, and Paragon Space Development Corporation was awarded $1.4 million to develop an Environmental Control and Life Support System (ECLSS) Engineering Development Unit designed to be used on different commercial crew vehicles. Sierra Nevada Corporation (SNC) received $20 million for the development of Dream Chaser, its reusable spaceplane, capable of transporting cargo and crew to LEO. The fifth company, United Launch Alliance (ULA), received $6.7 million for an Emergency Detection System (EDS) for human-rating its Evolved Expendable Launch Vehicle (EELV).

On April 18, 2011, NASA announced it would award nearly $270 million to four companies as they met CCDev 2 objectives. These objectives included: the capability of a vehicle to deliver and return four crew members and their equipment; to provide crew return in the event of an emergency; to serve as a 24-hour safe-haven; and to remain docked to the ISS for 210 days. Winners of funding in the second round of the CCDev Program included Blue Origin, which was awarded $22 million to develop advanced technologies in support of its orbital vehicle, including launch abort systems and restartable hydrolox (liquid hydrogen/liquid oxygen) engines. SNC received $80 million to develop phase 2 extensions of its lifting body-inspired Dream Chaser spaceplane. SpaceX was awarded $75 million to develop an integrated launch abort system design for its Dragon spacecraft. The system, reputed to have advantages over the more traditional tractor tower approaches used on prior manned space capsules, would be part of the company's Draco maneuvering system, currently used on the Dragon capsule for in-orbit maneuvering and de-orbit burns. Industry juggernaut Boeing proposed additional development for its seven-person CST-100 spacecraft, beyond the objectives for the $18 million received from NASA in CCDev 1. Designed to be used up to ten times, the capsule would have crew and cargo configurations, and was designed to be launched by different rockets.

While not selected for funding, ULA's proposed development work to human-rate the Atlas V rocket persuaded NASA to enter into an unfunded SAA with the company. A similar agreement was made with ATK and Astrium, which had proposed development of the Liberty rocket derived from Ares I and Ariane 5. A third unfunded agreement was made between the agency and Excalibur Almaz Inc. (EAI), which had proposed developing a crewed system incorporating modernized Soviet-era space hardware designs intended for tourism flights to orbit. EAI's concept for Commercial Crew to the ISS was to use the company's three-person space vehicle with an intermediate stage and fly the integrated vehicle on a commercially available launch vehicle.

In common with many government funding processes, to the outsider, NASA's rationale for CCDev awards seemed difficult to grasp. Consider the following proposals. One company proposed continuing work on a project that NASA had already funded, to human-rate a pair of highly reliable rockets that at least three companies wanted to use to launch their commercial spacecraft. Another company requested funds to build a new booster that had never flown and which no one intended to use. Which one did NASA fund? Neither. Such is the perplexing world of NASA contract awards. The companies were ATK, which proposed its Liberty rocket, and ULA, which was developing technologies to human rate its Atlas V and Delta IV launchers. Why did neither receive funding? Surprisingly, at least part of the reason seemed to have had little to do with the quality of the proposals. The rationale went something like this: spacecraft proposals were weighted higher than those for launch vehicles, for the simple reason that American companies have considerable experience developing launch vehicles, but at the time of the awards no U.S. company had successfully developed a crew-carrying spacecraft – at least not in the last 30 years. Given this emphasis, it is easier to understand why Boeing and SNC received funds to develop their human vehicles and why SpaceX received funding to human rate its Dragon spacecraft (Crew Dragon – Figure 2.8) and Falcon 9 rocket, the development of which NASA had been funding under the COTS program. Blue Origin was also developing a human-rated vehicle, so it received money to develop its biconic capsule and reusable rocket. Meanwhile, proposals that focused solely on rocket development received nothing.

Figure 2.8. Crew Dragon *Endeavour* approaches the ISS on April 24, 2021. Credit NASA

While the experience of developing launch vehicles versus the experience of developing spacecraft provided part of the justification when it came to awarding funding, the strengths and weaknesses of each program were also factored into the decision-making, so it is worthwhile looking at how NASA makes these assessments. Two factors NASA is particularly interested in are a company's Technical Approach and Business Information. These factors are color-coded, with green indicating a High Level of Confidence and white indicating a Moderate Level of Confidence. For example, ULA, which had sought $40 million in CCDev 2 funding, rated surprisingly low. NASA's reviewers identified several strengths in ULA's proposal, including the company's use of existing flight proven vehicles and infrastructure, its adaptable emergency detection system, a strong performance capability for crew abort scenarios, and an effective and integrated organizational structure. Its business information also did not fare too badly, since it was deemed suitable to deliver the proposed capabilities, it had a strong, highly experienced management team, possessed the requisite facilities, and had experienced and knowledgeable suppliers. With such a strong resume, you may be wondering why the company was not funded, but the reviewers also found weaknesses. These included a lack of definition of a critical path to an initial launch capability and correlation to CCDev 2 efforts, and a failure to adequately describe the commercial market to which it would provide products and services.

Ultimately, NASA deemed that ULA's work on their existing launch vehicles was not on the critical path for any crew transportation system and therefore the company did not accelerate the availability of crew transportation capabilities. This, after all, was a primary goal of the funding announcement. Nevertheless, while this assessment was true, it seemed strange that NASA would deny funding to a company that was developing a rocket which two CCDev 2 funded companies (SNC and Blue Origin) had hoped to use. Both had stated that they wanted to use ULA's Atlas V rocket to launch their crew vehicles (Blue Origin eventually shifted to its own reusable launcher). Boeing, which also received CCDev 2 funds, had also expressed their intention to use Atlas V to launch its CST-100 spacecraft, although the vehicle is being designed for multiple launchers.

Another strange case was the non-funding of ATK's Liberty rocket, at least at first glance. This launcher comprised a first stage developed from the canceled Shuttle-derived Ares I booster and the second stage of Europe's Ariane 5. From the NASA reviewer's perspective, while these technologies each had excellent flight heritage, the problem was that no one had committed to flying on the rocket. You can understand the agency's concern. NASA could have funded the Liberty all the way through the development phase but there would always be the risk that no spacecraft developer would select the launch vehicle as part of its design. Another black mark against ATK was their failure to provide NASA with sufficient details to assess launch vehicle environments (such as staging and abort scenarios) on the company's proposed upper stage or at the crewed spacecraft interface. While the company provided a solid technical approach, their details on environments did not provide NASA with enough confidence to accelerate this launch vehicle for use with the variety of crewed spacecraft. So, rather than use limited CCDev 2 funds on a launch vehicle with a questionable technical approach, the agency instead decided to select an extra spacecraft.

On August 3, 2012, NASA announced awards that had been made to three American commercial companies under yet another funding program – CCiCap (Commercial Crew Integrated Capability). Advances made by these companies under the signed SAAs through the agency's CCiCap initiative were intended to lead to the availability of commercial human spaceflight services for government and commercial customers. The CCiCap partners were SNC, which received $212.5 million, SpaceX, which received $440 million, and the Boeing Company, which received $460 million. As an initiative of NASA's Commercial Crew Program (CCP), CCiCap was an administration priority. The objective of the CCP (see Appendix IV) was to facilitate the development of a U.S. commercial crew space transportation capability, with the goal of achieving safe, reliable and cost-effective access to and from the ISS and LEO. After the capability was matured and expected to be available to the government and other customers, NASA planned to contract further, to purchase commercial services to meet its station crew transportation needs.

Shortly after CCiCap, the winners of the first phase of the Certification Products Contract (CPC) were announced. This first phase of the CPC reviewed the crew transportation systems, focusing particularly on engineering standards and designs of spacecraft systems. Winners of this phase were announced on December 10, 2012, and included $10 million being awarded to SNC, $9.6 million to SpaceX, and $9.9 million to Boeing. The second phase of the CPC, which involved an open competition that included testing and verification of crewed demonstration flights, was dubbed Commercial Crew Transportation Capability (CCtCap). Contract funding for CCtCap (see sidebar: *CCtCap*) occurred in 2014, a year when the program's budget was $696 million.

CCtCap

The CCtCap contract was the second phase of a two-phase procurement strategy to develop a U.S. commercial crew space transportation capability, to achieve safe, reliable and cost-effective access to and from the ISS with a goal of no later than 2017. Performance-based payments were to be used in this competitive, negotiated acquisition. Under CCtCap the final Design, Development, Test, and Evaluation (DDTE) activities necessary to achieve NASA's certification of a Crew Transportation System were conducted. The contract would be issued under Federal Acquisition Regulations Part 15 and would be Firm Fixed Price (FFP).

To comply with the award, each company had to perform at least one crewed test flight to prove their spacecraft could dock with the ISS. If everything checked out, NASA would award up to six crewed crew rotation flights to the ISS. In September 2014, NASA announced that Boeing and SpaceX had received contracts, with Boeing receiving $4.2 billion and SpaceX receiving $2.6 billion (the contracts included two operational flights). SpaceX would fly its Crew Dragon on its Falcon 9, while Boeing would fly its CST-100 Starliner using United Launch Alliance's Atlas V. SNC missed out on this award, so they did what any company does when it loses a government contract and submitted a protest to the Government Accountability Office (GAO). Unfortunately for SNC, their protest was denied in January 2015.

"SpaceX is deeply honored by the trust NASA has placed in us. We welcome today's decision and the mission it advances with gratitude and seriousness of purpose. It is a vital step in a journey that will ultimately take us to the stars and make humanity a multi-planet species."

Elon Musk

So, SpaceX and Boeing went ahead and started work developing their spacecraft for CCtCap (Table 2.4), while SNC announced they would downgrade their Dream Chaser to the Dream Chaser Cargo System (DCCS). For Boeing, the CCtCap announcement represented more money in the coffers, in addition to the already significant funding the multi-national company had received for its role as prime contractor for the SLS.

Table 2.4: Development Missions for SpaceX and Boeing

Mission	Spacecraft	Description	Crew	Date	Outcome
Dragon 2 Pad Abort Test	Dragon 2	Pad abort test	N/A	May 6, 2015	Success
Crew Dragon Demo-1	Dragon 2 C204	Uncrewed test flight. DM-1 launched on March 2, 2019 and docked to ISS. The Dragon spent 5 days docked to ISS before landing on March 8, 2019.	N/A	March 2, 2019	Success
Boeing Pad Abort Test	CST-100 Starliner	Uncrewed Pad Abort Test	N/A	November 4, 2019	Success
Boe-OFT	CST-100 Starliner	Uncrewed test flight. Originally planned to spend 8 days docked to ISS. Starliner was unable to rendezvous with ISS due to MET anomaly forcing it to enter lower-than-expected orbit. Spacecraft returned on December 22, 2019 after 2 days in orbit.	N/A	December 20, 2019	Partial failure due to MET anomaly
Crew Dragon In-Flight Abort Test	Dragon 2	A Falcon 9 booster launched a Dragon 2 capsule to test launch abort system. Abort occurred at 84 seconds after launch. Dragon 2 successfully separated from the Falcon 9 and flew away using its SuperDraco thrusters.	N/A	January 19, 2020	Success
Crew Dragon Demo-2	Dragon 2	Crewed test flight. Dragon 2 launched with two crew members and docked to ISS.	Doug Hurley Bob Behnken	May 30, 2020	Success
Boe-OFT 2	CST-100 Starliner	Uncrewed test flight. Suggested by Boeing and approved by NASA on April 6, 2020, due to the partial failure of software on previous Starliner test flight.	N/A	Early 2022	Planned

The outcome of the CCtCap CCP was the Crew Dragon, a crewed vehicle capable of ferrying astronauts to the ISS. In November 2019, NASA's inspector general published an audit report listing the seat prices for Crew Dragon at $55 million per seat, and $90 million per seat for the Starliner (compared with $86 million per seat on the Soyuz).

3

The Engines

"The meek shall inherit the Earth – the rest of us will go to the stars."

Robert A. Heinlein

Figure 3.0. The SuperDraco is used by the SpaceX Cargo Dragon spacecraft to maneuver in orbit and during re-entry. SuperDracos are used on Crew Dragon spacecraft as part of the vehicle's launch escape system. Credit SpaceX.

© Springer Nature Switzerland AG 2022
E. Seedhouse, *SpaceX*, Springer Praxis Books,
https://doi.org/10.1007/978-3-030-99181-4_3

In May 2021, Elon Musk informed the media that SpaceX had the capability to produce a Raptor rocket engine every 48 hours. The Raptor, described later in this chapter, is an interplanetary rocket engine capable of generating 500,000 pounds of thrust. In the SpaceX interplanetary colonization plans, it is the Raptor (Figure 3.1) that will provide the necessary power to send the Starship on its way. But why does there have to be such a quick production rate? Unlike NASA, which has also set its sights on a manned Mars mission, the Mars Messiah's goal is to establish a self-sustaining colony on the Red Planet, and such colonies cannot be achieved with small spacecraft carrying four or five astronauts. No, in the Musk universe, such colonies will require building a thousand Starships at a rate of 100 Starships per year. Given that the Starship and the Super Heavy Booster will be sporting 39 Raptors, that means a lot of Raptors must be built every year (Musk reckons SpaceX will need 800 to 1,000 engines per year), hence the requirement for such an aggressive development rate. Understanding how SpaceX has achieved this level of development is the purpose of this chapter.

Figure 3.1. Raptor engine. Credit Brandon de Young

In its first decade, SpaceX developed a family of liquid-propellant engines that included the Kestrel, the Merlin-1, and the Draco and SuperDraco. These engines had been designed for the workhorse launch vehicles of the company, namely the Falcon 1, Falcon 9 and Falcon Heavy, in addition to providing thrust for the Dragon spacecraft. In November 2012, SpaceX announced that it planned to develop methane-based engines using staged cycle combustion for higher efficiency. As with most aspects of SpaceX, the development path was not exactly… traditional. For instance, in many commercial ventures it is not unusual to develop a product without developing an engine to power it. This is because there is a ready availability of engines in different power ranges, which allows companies to produce an array of machines almost as quickly as they can be imagined. Even in highly specialized applications, where the final product is built in limited production numbers, the power plant usually remains common almost to the point of being mundane. This is not the case when it comes to space launch systems.

Until quite recently, the development of launch vehicles was characterized by implementing vehicle-specific engines which consumed a significant part of the rocket's development budget. But, in the last decade or so, rocket designers have begun to source common, off the shelf (Commercial Orbital Transportation Services, or COTS) solutions. For example, when Lockheed Martin was looking for an engine to power its Atlas V booster, it decided to use the Russian RD-180. Another example is Orbital Science Corporation (OSC), which decided to use refurbished Russian NK-33 engines to boost its Taurus 2 launch vehicle. The choice of the NK-33 was a surprise to many in the commercial spaceflight industry since these engines had a less than reliable reputation, having been the engine of choice for the failed Soviet N-1 booster program. The termination of that program opened the door for the engines being sequestered by OSC. Then there is the case of SpaceX, which could have gone down the same route as OSC or Lockheed Martin, but instead chose to develop its engines in-house, beginning with the Merlin. At 512,000 newtons (115,000 lbf.) sea level thrust, the diminutive Merlin (Figure 3.2) is perfect for second or orbital stage applications. It is a relatively simple engine by design, which runs on liquid oxygen (LOX) and rocket grade kerosene in a combustion process that utilizes a straightforward open cycle process, as opposed to staged[1].

---

[1] In an open cycle process, some of the propellant is burned in a gas-generator and the resulting hot gas is used to power the engine's pumps. The gas is then exhausted, hence the term 'open cycle'. In the staged combustion cycle, some of the propellant is burned in a pre-burner and the resulting hot gas is used to power the engine's turbines and pumps. The exhausted gas is then injected into the main combustion chamber, along with the rest of the propellant, and combustion is completed.

Figure 3.2.   Merlin engine. Credit Steve Jurvetson.

Before we discuss how the Merlin was developed, we must mention who developed it. As the chief designer of SpaceX, Elon Musk receives the lion's share of media attention, but some credit must go to the engineer who designed the engines. That engineer is Tom Mueller, who we mentioned in Chapter 1. One of SpaceX's first employees, Mueller now works as the company's Chief Technology Officer (CTO) for propulsion. Since being hired, he has picked up a Space Pioneer Award for his work on developing SpaceX engines. It was Mueller who brought his TRW experience to bear on the nation's first new, large, liquid-fuel rocket engine in 40 years. Unlike more complex engines that mix fuel and oxidizers at multiple points, the Merlin uses a single injector. This injector, which draws upon a long heritage of space-proven engines, is the heart of the Merlin engine. The injector is the pintle type first used in the Apollo Program for the Lunar Module Descent Engine (LMDE) that powered the Lunar Excursion Module (LEM). If you are wondering how a pintle injector works, think about your garden hose for a moment. At the end of the hose is a nozzle. Turn it one way and you get a steady narrow stream of water shooting out in a long arc; turn it to shut it off and you get a cone-shaped fan spray. When you look into the nozzle you will see a round pintle that moves back and forth as you turn the outer casing one way and the other. The fan-shaped water spray is what fuel and oxidizer spray looks like inside the rocket engine, the only difference being that the spray of water from your hose does not combust.

In 2000, TRW demonstrated a newer design of engine that used a pintle injector – the TR-106, also called the Low Cost Pintle Engine (LCPE). The LCPE generated 650,000 pounds of thrust, which was more than the 400,000 pounds of thrust generated by the Space Shuttle Main Engine (SSME). One of the largest liquid rocket engines built since the Saturn F-1 engines powered Apollo, the LCPE was designed as a simple, easy to manufacture, low-cost engine, constructed from common steel alloys using standard fabrication techniques. Instead of using expensive regenerative cooling, the LCPE employed ablative cooling techniques and featured the least complex type of rocket propellant injector – a single element coaxial pintle injector.

The LCPE's design was born of TRW's goal to design an engine that minimized cost while retaining excellent performance, rather than following traditional rocket design that sought maximum performance and minimum weight. By doing this, TRW hoped to reduce the cost of launch vehicles and enable access to space for government and commercial customers. The LCPE was subjected to hot fire testing at 100 percent of its rated thrust, as well as in a 65 percent throttle condition at NASA's John C. Stennis Space Center (SSC) in Mississippi. In between tests, TRW changed the pintle injector configuration three times to investigate the engine's performance envelope. Throughout the tests, the engine demonstrated rock solid performance, stability, and versatility. Since the engine provided the inspiration for the Merlin design, it is worth looking at some of the LCPE's features.

Perhaps one of the engine's signature features was its scalability. The LCPE was scalable over a range of thrust levels and propellant combinations, enabling it to be easily adapted to several launch vehicles capable of lifting anything from 200 to 200,000 pounds to Low Earth Orbit (LEO). The LCPE could also power the first stage of an Evolved Expendable Launch Vehicle (EELV) multi-stage launch vehicle, and scaled down versions of the engine could be used for the vehicle's second stage. Another of the LCPEs positive traits was *combustion stability*, meaning it could function over a wide range of operating conditions thanks to the unique injection and combustion flow fields created by the pintle injector. It was the pintle injector that was the key to many of the LCPE's performance attributes. Simple in design, the pintle injector contained only five parts, yet it was this element that permitted the LCPE's deep throttle capability. Given the LCPE's sterling performance and the huge strides TRW had taken toward providing more affordable access to space, it was a surprise to many that NASA cancelled further work on the engine. Less surprising was Tom Mueller's decision to go with a single-pintle design for the SpaceX Merlin rocket engines. Merlin's design was driven by the

Falcon 1 rocket (Figure 3.3). The goal was to design an engine that could withstand a 160-second burn, which would be long enough for the first stage of the Falcon 1 to reach an altitude of 90 kilometers. At this altitude, the second stage's smaller engine would kick in to boost the cargo to 130 kilometers and LEO.

Figure 3.3. The Falcon 1 was an expendable launch vehicle that achieved orbit in September 2008, powered by a single Merlin engine on its first stage and a Kestrel engine on its second. Credit SpaceX/ public domain.

Designing a rocket engine from scratch has never been an easy task, but in October 2003 Mueller and his engineers figured they were ready to fire up a Merlin on a test stand. In this first Merlin variant, propellant was fed via a single shaft, dual impeller turbo-pump operating on a gas generator cycle. The turbo-pump also provided the high-pressure kerosene for the hydraulic actuators, which recycled into the low-pressure inlet, a design feature that eliminated the need for a separate hydraulic power system. This design also meant that thrust vector control failure caused by running out of hydraulic fluid was not possible. Another use of the turbo-pump was to provide roll control by actuating the turbine exhaust nozzle. Combining the three functions into one device allowed engineers to know that all systems were functioning before the vehicle lifted off, and resulted in a significant improvement in system level reliability. Unfortunately, the test did not go well. During the run, the Merlin's exhaust began to melt the metal in the engine's throat and the intense heat endangered the seals responsible for governing the propellant. After 60 seconds, the engineers decided to shut it down; any longer and the engine might have blown up.

Over the next 15 months, Mueller and his team of engineers worked to trouble-shoot the bugs. One of the fixes was to reduce the amount of LOX entering the injector, a solution that made the engine run cooler and strengthened the seals. To prevent more heat damage, Mueller treated the Merlin's nozzles with resin impregnated with silicon fibers, an ablative coating designed to char and flake off while the engine was running, taking damaging heat with it. Once the bugs were fixed, the engineers once again took their places in the bunker for the second test. This was planned to be a full mission duty cycle, equating to the time needed to deliver a payload into LEO.

The test proceeded without a hitch, with the Merlin shutting down after 162.2 seconds. It was an impressive performance, not just in terms of how well the Merlin performed but also because the Merlin was the first rocket engine to be developed in the United States since Rocketdyne's RS-68 engine (for Boeing's Delta IV) in the early 1990s, and only the second since the late 1970s when Rocketdyne developed the SSME. Now all Mueller's team had to do was to mount the engine on a Falcon 1 and launch it. As described in Chapter 1, the first launch was doomed before ignition due to the salty Pacific air that corroded an aluminum nut on the engine, causing a leak. The result was catastrophic. The spilled fuel caught fire and, 34 seconds after launch, flames burned through a pneumatic line and shut down the engine, causing the rocket to crash into the Pacific seconds later. A year later, sloshing fuel in the second stage of another Falcon 1 caused the rocket to spin out of control before reaching orbit. Then, during the third flight in August 2008, the first stage collided with the second stage shortly after separation.

Mueller's team persevered and, less than two months later, another Falcon I roared from its South Pacific launch pad. After a week spent reviewing data, Mueller confirmed that the flight had gone well. It was the culmination of six years of hard work by a very talented team and came as a relief to Musk who, two weeks before the successful flight, had received the American Institute of Aeronautics and Astronautics (AIAA) George M. Low Space Transportation Award for the most outstanding contribution to the field of space transportation[2]. The Falcon 1's fourth flight achieved orbit, with the first burn terminating at

---

[2] Established in 1988, the AIAA's George M. Low Space Transportation Award honors the achievements in space transportation made by Dr. George M. Low, who played a leading role in planning and executing the Apollo missions. The biennial award is presented for a timely outstanding contribution to the field of space transportation. Musk's citation read: For outstanding contribution to the development of commercial space transportation systems using innovative low-cost approaches.

330.5 km altitude and 8.99° inclination, which was pretty close to the intended insertion of 330 km altitude and 9.0° inclination. The second burn tested the restart capability[3], allowing the upper stage to coast for 43 minutes and boosting the orbit to 621 km. Incidentally, if you have ever wondered how SpaceX engines got their names, here is the explanation:

> *"When we first started SpaceX, we just called our booster engine the 60 K engine, but after we started running it Elon told me to come up with a name for it that wasn't numbers and letters (like RD-180, RS-68, etc.). One of the people working on the turbo-pump from Barber Nichols was a falconer and she suggested we name it after a falcon. I thought that sounded good, so I asked her what are some falcon names. She named off a bunch and I can't recall them all, but I do remember that the Kestrel is the small one, the Merlin is a medium size falcon, and the Peregrine and Gyrfalcon are large falcons. I thought great, we'll name the small second stage engine Kestrel and the medium sized engine the Merlin. I knew we would develop bigger engines in the future, so I planned to reserve Peregrine for later. Elon liked the naming, so it stayed. Years later we started work on a staged combustion engine which was a different type than Merlin, so I was thinking along the lines of Eagle or something. I eventually came up with Raptor, which is a general definition of birds of prey including Eagles, Hawks, Falcons and Owls. No, it's not named after a dinosaur! That was accepted as the name of the engines for BFR."*

Tom Mueller being interviewed in Quora.

Before we look at the Merlin variants, we must mention the Kestrel, another rocket engine SpaceX built around the pintle architecture. The Kestrel (which is no longer manufactured) was a high efficiency, low pressure vacuum engine that featured a vacuum thrust of 6,245 lbf and a vacuum specific impulse of 325 seconds. Developed by SpaceX to propel the upper stage of the Falcon 1, the Kestrel did not have a turbo-pump and was a pressure-fed rocket. It was ablatively-cooled in the chamber and throat, and radiatively cooled in the niobium nozzle, which was designed to be super strong at extreme temperatures and is highly resilient to impact. Thrust vector control was achieved by electro-mechanical actuators on the engine dome for pitch and yaw, while roll and attitude control during the coast phases was provided by helium cold-gas thrusters. A multiple restart capability on the upper stage was provided by a highly reliable triethyl-aluminum-triethylborane (TEA-TEB) pyrophoric system. TEA is a volatile, highly pyrophoric compound that ignites immediately upon exposure to air. Since TEA is one of the few

---

[3] In a multi-manifested mission, a restart capability allows delivery of separate payloads to different altitudes and inclinations.

substances pyrophoric enough to ignite on contact with cryogenic liquid oxygen (LOX), the substance is especially desirable as a rocket engine ignitor. TEB, TEA's cousin, which is also strongly pyrophoric, has a long history in the aerospace world, having been used to ignite the JP-7 fuel in the Pratt & Whitney J58 engines that powered the SR-71 Blackbird spy plane.

For those who have followed the SpaceX story, the Merlin's development can be a little confusing because there have been several variants, some of which have flown and some which have not. The initial version, the Merlin 1A, used an expendable, ablatively-cooled carbon fiber composite nozzle, and produced 340 kN (77,000 lbf) of thrust. This variant flew only twice, the first on March 24, 2006, when the engine caught fire and failed due to the fuel leak shortly after launch, and the second time on March 21, 2007, when it performed successfully.

The Merlin 1B rocket engine was an upgraded version of the Merlin 1A. Capable of producing 380 kN (85,000 lbf) of thrust thanks to a turbine upgrade, the initial use of the Merlin 1B was intended for the Falcon 9 launch vehicle, which would have housed a cluster of nine (hence the designation) Merlin 1Bs on the first stage. However, based on experience from the Falcon 1's first flight, the 1B was never used on a flight vehicle and SpaceX moved its Merlin development to the Merlin 1C. The Merlin 1C uses a regeneratively-cooled[4] nozzle and combustion chamber. It was this engine that powered the successful fourth Falcon 1 flight in September 2008 (it also powered the third unsuccessful flight), and the Falcon 9 on its maiden flight in June 2010. When configured for Falcon 1 vehicles, the Merlin 1C had a sea level thrust of 350 kN (78,000 lbf) and a vacuum specific impulse of 304 seconds, but when configured for the Falcon 9 the sea level thrust is 560 kN and the specific impulse is 300 seconds.

The follow-up to the Merlin 1C was the Merlin Vacuum (Figure 3.4), which featured a larger exhaust section and a significantly larger expansion nozzle to maximize the engine's efficiency in the vacuum of space. In common with its predecessor, the Merlin Vacuum's combustion chamber was regeneratively-cooled, while the niobium alloy expansion nozzle was radiatively-cooled. The engine produced a vacuum thrust of 411 kN and a vacuum specific impulse of 342 seconds. On January 2, 2010, the first production Merlin Vacuum engine underwent a full duration orbital insertion firing lasting 329 seconds, before being flown on the second stage for the inaugural Falcon 9 flight on June 4, 2010.

---

[4] As you can imagine, firing a rocket generates tremendous heat, with combustion temperatures reaching 2,500 to 3,600 K. To cool the thrust chamber, rocket designers use a variety of chamber cooling techniques, the most widely used of which is *regenerative cooling*. This method cools the thrust chamber by flowing high-velocity coolant over the back side of the chamber's hot gas wall to convectively cool the hot gas liner. The coolant with the heat input from cooling the liner is then discharged into the injector and utilized as a propellant.

Figure 3.4. Merlin 1C Vacuum engine. Credit SpaceX/public domain.

An unplanned test of a modified Merlin Vacuum engine took place in December 2010. Shortly before the second flight of the Falcon 9, two cracks were discovered in the engine's niobium alloy sheet nozzle. Engineers decided to cut off the lower 1.2 meters of the nozzle and launch two days later, since the extra performance that would have been gained from the longer nozzle was not necessary to meet the mission objectives. Even with the shortened nozzle, the engine still placed the second stage into an orbit of 11,000 kilometers.

In June 2012, SpaceX further expanded its stable of rocket engines by successfully test firing the Merlin 1D engine (Figure 3.5). Designed to propel future SpaceX vehicles into LEO, the 1D was the follow-up to the Merlin engines used to ferry the Dragon to the ISS the previous month. Thanks to an improved thrust-to-weight ratio, the Merlin 1D is the most efficient booster engine ever built, featuring a vacuum thrust of 690 kN (155,000 lbf) and a vacuum specific impulse of 310 seconds. In addition to sporting impressive propulsion numbers, the new Merlin also featured a throttle capability from 100 percent to 70 percent and a 160:1 thrust-to-weight ratio while still maintaining the structural and thermal safety margins needed to carry astronauts, the highest ever achieved for a rocket engine. The engine achieved its first full mission duration firing in June 2012, in a test that featured multiple restarts at a target thrust and specific impulse. The engine firing was for 185 seconds with 147,000 pounds of thrust, which was the duration and power required for a Falcon 9 launch. The Merlin 1D achieved flight qualification in March 2013, and three months later the engine completed development testing when the 1D was used on a Falcon 9 1.1 first stage, on a flight that launched a Canadian Space Agency satellite. Later that year, Musk announced the engine was operating at 85 percent of its thrust and the company expected to increase this on future iterations of the Falcon 9. In June 2015, the Merlin 1D had its thrust upgraded to 162,500 lbf, while in May 2018 this figure had improved to 190,000 lbf, a thrust increase that improved the Falcon 9's payload capability to 22 metric tons to LEO.

Figure 3.5. Test firing of the Merlin 1D engine. Credit SpaceX.

There is no doubt that the Merlin 1D engines (see sidebar: *Merlin Legacy*) have made the next generation of Falcon rockets a lot more powerful, but where do these engines fit in the realm of space exploration? Nine Merlin 1C engines, the standard configuration for the Falcon 9's first stage, generate about 1.1 million pounds of thrust at launch, whereas the Merlin 1D engine's 185-second test fire in June 2012 delivered 147,000 pounds of thrust. Nine of these new engines clustered on the Falcon 9's first stage will give the rocket nearly 1.5 million pounds of thrust at launch. Applied to the Falcon Heavy (Figure 3.6), which used 27 engines in its first stage, these upgraded Merlin engines combined to generate more than 5 million pounds of thrust at liftoff (about the same as 18 Boeing 747 aircraft for those who like these sort of comparisons). Further development of the Merlin 1D has resulted in even more impressive numbers. In May 2018, SpaceX announced 190,000 pounds of thrust had been achieved, a number which is close to the sea level thrust of the engines used on the Saturn 1 and Delta II.

Figure 3.6. Falcon Heavy demonstration mission, February 6, 2018. The launch vehicle comprised three Falcon 9 nine-engine cores. The combined thrust of the 27 engines was approximately 5 million pounds. Credit SpaceX.

Merlin Legacy

Why has the Merlin design been so successful? The engine is a manageable size and, as far as rocket designs go, the system is also relatively basic since the Merlin runs on arguably the simplest combination of fuels – LOX and rocket grade kerosene – and utilizes a straightforward open cycle (as opposed to staged) combustion process. Compared to liquid hydrogen-based systems, RP-1 offers an advantage in terms of ease of system design and operational handling. It is this simplicity that confers a reduced cost, and less complexity in terms of propellant piping, seals, valves, and insulation, as well as the ability to incorporate smaller fuel tanks. By selecting just one basic engine to be used for its family of launch vehicles (except for the Falcon 1 second stage Kestrel engine), SpaceX has not only been able to reduce design and production costs, but also to allow engineers to acquire a rapid buildup in experience (after all, ten engines are utilized in each Falcon 9 flight). With the company's busy manifest over the years, SpaceX has already flown several hundred Merlin engines, gathering experience that has provided plenty of opportunity to examine and refine its engines post-flight with an eye towards future improvements. Additionally, given the number of cargo missions slated to fly and the potential for the engine to play a role outside the Falcon family of launch vehicles, it is very likely that the Merlin will be destined for an exceptionally long life span.

How does the Merlin 1D compare with legacy engines? Well, consider the F-1. Developed by Rocketdyne, the F-1 was used in the S-IC first stage of each Saturn V, which served as the main launch vehicle in the Apollo program. It is still the most powerful single-chamber liquid-fueled rocket engine ever developed. During launch, this behemoth burned 1,789 kg of LOX and 788 kg of RP-1 *every second*, generating 6.7 million newtons of thrust. During their two-and-a-half minutes of operation, the cluster of F-1s propelled the Saturn V vehicle to 68 km altitude and a speed of 9,920 km/h. Each F-1 engine had more thrust than three SSMEs (no slouch in the propulsion department) combined. Despite being based on 45-year-old technology, the F-1 engines have not been forgotten; in 2012, it was proposed to use the engines to launch the Space Launch System (SLS) rocket that NASA hopes to use to send its astronauts to a Deep Space Gateway, the Moon and perhaps Mars.

The Merlin engine iteration did not stop at the Merlin 1D. At the AIAA Joint Propulsion conference on July 30, 2010, McGregor rocket development facility director Tom Markusic shared information with the media about the development of a new SpaceX engine, the Merlin 2. Designed at the time to launch conceptual super-heavy-lift launch vehicles, the Merlin 2 was intended to be an LOX/RP-1-fueled engine, capable of a projected 7,600 kN (1,700,000 lbf) of thrust at sea level and 8,500 kN (1,920,000 lbf) in a vacuum. According to the presentation at AIAA, SpaceX reckoned the engine could have been qualified in three years for $1 billion. That idea was short-lived however, as SpaceX soon began focusing more on the development of its methane-fueled engine, dubbed Raptor. The Raptor, which we will discuss shortly, will be the workhorse of the SpaceX hardware for Mars, the much-hyped Mars Starship. Before we get to the Raptor, we need to mention the SpaceX family of smaller rocket engines. On a smaller scale than the Merlin family is the Draco (Figure 3.7), a hypergolic rocket engine/thruster designed for use on the Dragon and the upper stage of the Falcon 9 rocket.

The Draco thrusters (four are used on the Falcon second stage) generate 400 newtons (90 lbf) of thrust using a mixture of monomethyl hydrazine fuel and nitrogen tetroxide oxidizer. Used for attitude control and maneuvering, the thrusters are dual-redundant in all axes, which means any two can fail and astronauts will still have vehicle control in pitch, yaw, roll and translation.

Figure 3.7. A Draco thruster fires as Dragon approaches the ISS during the COTS 2 mission. Credit NASA.

An addition to the Draco family was announced by SpaceX on February 1, 2012. The company told the media it had also completed development of a more powerful version of the Draco thruster called, appropriately enough, the SuperDraco (Figure 3.8). A throttleable engine with multiple restart capability, the SuperDraco is designed for the Launch Abort System (LAS) on the SpaceX Dragon spacecraft. In the event of a launch abort, eight SuperDracos will fire for five seconds at full thrust. The engine, whose development was partially funded by NASA's CCDev 2 program, generates a thrust of 67,000 newtons (15,000 lbf), making it the second most powerful engine developed by SpaceX. It is more than 170 times more powerful than the Draco (and twice as powerful as the Kestrel) and about 1/9th of the thrust of a Merlin 1D engine. The first firing of the SuperDracos occurred on May 6, 2015 during a Crew Dragon Pad Abort Test. In addition to using the SuperDraco thrusters for the LAS, SpaceX also considered plans to use them for powered landings on Earth. However, that plan failed to meet NASA's exacting safety standards, so SpaceX had to revert to landing under canopy.

And so to the Raptor engine, which was first mentioned by Max Vozoff at the AIAA Commercial Crew/Cargo Symposium in 2009, although details were sketchy. A staged combustion engine powered by cryogenic liquid methane and

Figure 3.8. SuperDraco test firing. Credit SpaceX.

liquid oxygen, the Raptor concept was finally revealed in 2013 as an engine that would be the enabler of manned flights to Mars. The Raptor is used on the SpaceX Starship (previously the Interplanetary Transport System). Up to 2015, Raptor engine development was funded solely by SpaceX, but in 2016 the USAF gave the company $33.6 million to develop a Raptor variant that could be used on an upper stage of a Falcon 9 or a Falcon Heavy. Shortly after the USAF announcement, a development Raptor was delivered to the SpaceX testing location in McGregor, Texas, where it was test fired in September 2016. Why is the Raptor fueled by methane? Simple. There is a lot of the gas on Mars. Bucket-loads in fact. Assuming that In-Situ Resource Utilization (ISRU) can be employed on Mars, space missions will become more affordable. In terms of the Raptor's performance, more details were announced in February 2014 at the *Exploring the Next Frontier: The Commercialization of Space is Lifting Off* event, when Tom Mueller revealed that the Raptor's intended use was on a vehicle (the Starship) capable of delivering 100 tons to Mars. To achieve this goal, the Raptor would produce more than 1,000,000 lbf. Four months later, Mueller upgraded this number to 1,600,000 lbf. The Raptor (Figures 3.9 and 3.10) was going to be a powerful engine.

Figure 3.9. Testing the Raptor at Stennis Space Center. Credit NASA.

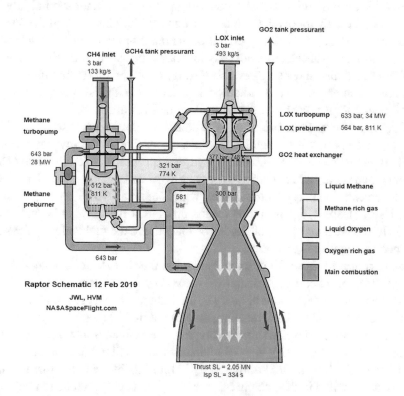

Figure 3.10. Raptor combustion scheme. Credit NASA.

SpaceX wasted no time in development testing. Hot-fire test followed hot-fire test, and in August 2016 SpaceX announced it had built the first scaled-down integrated Raptor, an engine sporting 220,000 lbf thrust. A little slice of history was made, because the baby Raptor was the first ever full-flow staged-combustion methane/oxygen engine to reach a test stand. Musk, the Raptor's proud father, tweeted images of the Raptor on September 26, 2016. Twelve months later, the scaled down Raptor had completed more than 40 tests and more than 1,000 seconds of firings. More tests followed, and in July 2019 the Raptor engine made its first flight on the Starhopper test vehicle. We will get to the flight tests shortly, but before we do it is worth mentioning some of the other features of this cutting-edge engine. At the time of writing, the Raptor holds the record for the highest combustion chamber pressure attained by an operational rocket engine (330 bar). For those rocket engineers among you, the Raptor operates using a full-flow staged-combustion cycle. For those of you who are not rocket engineers, this means that oxidizer and fuel are mixed in the gas phase before entering the combustion chamber. Why this design? It is all about increasing performance and/or reliability, but those are not the only innovative features of this engine. There is also the way the engines are constructed, with 40 percent of the mass of the engine manufactured by 3-D printing. Obviously, this not only increases the speed of development and manufacture, but also speeds up iterative testing. Together, this streamlined operation results in an engine that is ridiculously affordable (for the spaceflight industry), with each Raptor costing only $1 million in 2019 and a forecast cost of just $250,000 once the engines are in mass production. Before we discuss the flight testing of the Raptor, it is instructive to take a closer look at the advantages of the propellant. Why does SpaceX use deep cryogenic propellant and what is it? Deep cryogenic means the fluid is cooled to very close to its freezing point. This is different to conventional cryogenic fluids, which operate closer to their boiling points. Using deep cryogenic fluids means propellant density can be increased, which in turn means more propellant can be loaded into the fuel tanks. At the same time, specific impulse is increased.

Now, to return to those flight tests. Following the Starhopper test (which was one year *ahead* of schedule incidentally – when do you ever hear of any test being ahead of schedule in the launch industry?), the Raptor made its second Starhopper flight test on August 27, 2019, reaching an altitude of 150 meters. A year later, a Raptor engine powered a Starship prototype to the same altitude. A few months after that, in December 2020, three Raptors powered a Starship to an altitude of 12.5 kilometers. All went well until shortly before landing, when a low methane feed pressure resulted in the vehicle being destroyed. The next launch in February 2021 resulted in another failure following a flight to 10 kilometers, an event that repeated itself in two subsequent launches in March 2021. Finally, in May 2021, a Starship prototype powered by three Raptor engines reached 10 kilometers altitude before successfully landing on the pad.

As the Raptor continues its flight testing, there are many who are interested in knowing how the Raptor stacks up against other engines on the market. That query is answered by the data presented in Table 3.1. But mere numbers do not even begin to tell the story of just how extraordinary an engine the Raptor is. The closest comparable engine is Blue Origin's BE-4, which began testing in October 2017, a full 16 months before testing began on the Raptor. But how many engines has Blue Origin built and tested? Nine – in four years. At the time of writing, SpaceX is building more than a dozen Raptors *per month*.

Table 3.1: Comparison of Raptor with other engines

Engine	Rockets	Thrust	Specific impulse, vacuum	Thrust to- weight ratio	Propellant	Cycle
Raptor sea-level	Starship	2,200 kN	~350 s	200 (goal)	LCH_4 / LOX	Full flow staged combustion
Raptor vacuum			~380 s	<120		
Merlin 1D sea-level	Falcon booster stage	914 kN	311 s	17	RP-1 / LOX (subcooled)	Gas generator
Merlin 1D vacuum	Falcon upper stage	934 kN	348 s	180		
Blue Origin BE-4	New Glenn, Vulcan	2,400 kN			LNG / LOX	Oxidizer-rich staged combustion
Energomash RD-180	Atlas III, Atlas V	4,152 kN	338 s	78.44		
Energomash RD-191/181	Angara, Antares	2,090 kN	337.5 s	89		
Kuznetsov NK-33	N1, Soyuz-2-1v	1,638 kN	331 s	136.66		
Rocketdyne RS-25	Space Shuttle, SLS	2,280 kN	453 s	73	LH_2 / LOX	Fuel-rich staged combustion
Rocketdyne F-1	Saturn V	7,740 kN	304 s	83	RP-1 / LOX	Gas generator

4

The Rise of the Falcon

"You came in that thing? You're braver than I thought."

Leia Organa, upon seeing the Millennium Falcon for the first time.

Figure 4.0. SES-10 Launch. Yet another SpaceX first: the first re-flight of an orbital-class rocket. Credit: SpaceX.

© Springer Nature Switzerland AG 2022
E. Seedhouse, *SpaceX*, Springer Praxis Books,
https://doi.org/10.1007/978-3-030-99181-4_4

In the sci-fi classic, *Star Wars*, the iconic *Millennium Falcon* spacecraft is commanded by Harrison Ford's character Han Solo. Elon Musk must have been a fan because the SpaceX fleet of Falcon rockets is named after George Lucas's fictional spaceship, although the two vehicles bear little resemblance and their roles have little in common. Whereas the *Millennium Falcon* was designed to go tearing around the galaxy looking for trouble, its namesake was designed to provide breakthrough advances in reliability, cost, and time to launch. And those advances have come thick and fast. In little more than a decade, SpaceX has developed a whole set of new technologies for orbital vehicles. These technologies have included full and rapid reusability, returning launch vehicle first stages to launch sites, routine re-flights, and supersonic retro-propulsion, the latter a key enabling technology for realizing Musk's goal of landing humans on Mars. Given that bold goal, it is not surprising there is so much attention given to the Starship system, but the Starship could not have existed without the Falcon, so what follows is an overview of the history of this launch vehicle and its operations. To do that we must turn the clock back to 2002, the year SpaceX was founded.

With funds stretched tight after founding SpaceX, Musk decided to develop a small practical space launcher. That launcher turned out to be 21.3 meters tall, 1.7 meters in diameter and featured a rocket-grade kerosene-fueled (RP-1) two-stage rocket capable of boosting about 0.6 tonnes to Low Earth Orbit (LEO). Its first stage was boosted by the in-house developed Merlin engine, while the second stage was powered by the Kestrel engine. The first stage, which was helium pressurized and designed to be recovered at sea, was a pressure-assisted, stabilized, graduated monocoque aluminum design that used a common bulkhead between its aft kerosene tank and its forward liquid oxygen (LOX) tank. The expendable second stage, which was also helium pressurized[1], was fabricated from aluminum; the plan had been to use lighter aluminum-lithium, but SpaceX was unable to secure the metal. The Falcon 1's mission architecture was as simple as it was elegant: a single SpaceX Merlin engine powered the Falcon 1 first stage and, after engine start, the vehicle was held down until all systems were verified to be functioning normally, at which point the vehicle blasted off. Stage separation was achieved via dual initiated separation bolts and a pneumatic pusher system, and the retrievable first stage returned by parachute to a water landing where it was recovered by ship in a procedure similar to the one employed to recover the Shuttle's Solid Rocket Boosters (SRBs). Incidentally, the SpaceX parachute recovery system is built by Airborne Systems Corporation, which also built the Shuttle booster recovery system.

---

[1] Helium tank pressurization is achieved by composite over-wrapped Inconel tanks manufactured by the Arde Corporation – the same model is used in Boeing's Delta IV rocket.

Since we have already discussed the events leading up to the first successful Falcon 1 launch in Chapter 1, there is no need to repeat those events here. Instead, to provide readers with an insight into the world of spaceflight operations, the following is an overview of what SpaceX provided to potential customers and how SpaceX executed operations. Understanding these processes is useful because most of the operations that supported the Falcon 1 are identical to those supporting the Falcon 9 today.

So, what would you do if you happened to be a company interested in launching a payload – a satellite for example – onboard the Falcon 1? One step would have been to consult and complete a payload questionnaire. This would allow SpaceX to assess mission feasibility, define mission requirements, and fine-tune launch countdown procedures. The questionnaire required the customer to provide detailed information about their payload, such as a mathematical model, which was needed before mission integration could be performed. In addition to the mathematical model, customers also had to provide an Interface Control Document (ICD) that described all mission-specific requirements. An environmental statement was also required, as was a specification of the radio frequencies that would be transmitted by the payload during the ground processing and launch operations. This latter prerequisite went into some detail, requiring the customer not only to list the individual frequencies but also the names and qualifications of the personnel who would operate the radio frequency systems, the duration of transmission, and the frequency bandwidths. Details of the payload design, including graphics, and configuration drawings showing dimensions were also required, together with the procedures that were planned at the launch site operations. This information was then passed on to government agencies and range safety.

Why all this information? It is because the SpaceX customer interface has not changed. Today, for those wanting to fly payloads on the Falcon 9 or Falcon Heavy, customers must first review the Payload User Guide (PUG) and follow the procedures outlined therein. The PUG is a publication not dissimilar to a car user's manual, which provides a detailed set of instructions concerning what can and cannot be accommodated. Since the steps involved in flying a payload on board a Falcon 9 are procedurally very similar to those used for flying on board a Falcon 1, we will revert to the present tense. In addition to the standard performance characteristics of the vehicle, the PUG also describes very specific information about payload integration, which informs potential customers about the options available to them when launching a payload. For many customers, it is important to understand the operating conditions that will affect their payload, not only during launch but also while the payload is in orbit. For example, a customer hoping to launch a payload that contains consumables requires a different set of operating conditions than one with no consumables.

In addition to perusing the PUG, customers might be interested in some of the standard services provided by SpaceX. For example, a Collision Avoidance

Maneuver (CAM) is provided as a standard service at no extra charge. The CAM – if required – can be performed using the Reaction Control System (RCS) thrusters, which are tilted forward 20 degrees and positioned to minimize gas impingement on the spacecraft while still providing adequate separation. Another service provided to customers is a restart capability, which provides the flexibility required for payload insertion into orbits with varying eccentricities, or the deployment of multiple payloads into different orbits. The Falcon 1's Guidance, Navigation and Control (GNC) system included a ruggedized flight computer and an Inertial Measurement Unit (IMU) backed by a Global Positioning System (GPS) receiver which was flown for navigation updates. The GNC system also included an S-Band telemetry system, an S-Band video downlink, and a C-Band transponder.

SpaceX provided a standard payload separation system for the Falcon 1, but the company could also integrate a separation system chosen and supplied by the payload provider. Payload separation, which was initiated non-explosively by separation springs that imparted separation velocity, was a timed event referenced to the second stage burnout. Other options provided to the customer included spinning up the payload, which could reach up to six revolutions per minute (rpm) at separation. Attitude was another option that could be customized, with attitude and rate accuracies that included plus or minus two degrees of roll, half a degree of pitch and yaw, and one tenth of a degree of body rate.

Another selling feature of the Falcon 1 was its reliability. SpaceX took heed of the lessons learned from previous launch vehicle failures (most of which were attributed to engines, avionics, or stage separation failures) and built a robust propulsion system with a redundant ignition system, matching this with a vehicle that featured a state-of-the-art avionics system. In fact, one of the primary goals when designing the Falcon 1 was to ensure reliability was not compromised, an approach that required some key choices. Among these was designing the first stage to be recovered and reused, which meant this part of the vehicle had to have significantly higher margins than an expendable stage. During testing, SpaceX subjected a first stage to more than 190 cryogenic pressure cycles, with no evidence of fatigue. Another decision that featured in the reliability matrix was to minimize risk during the propulsion and separation phase, a goal achieved by designing the Falcon 1 with the minimum number of engines in serial. Off-nominal propulsion events were further reduced by the company's launch operations procedures, which involved holding down the first stage after ignition – but prior to release – to observe engine trends; if an off-nominal condition was detected, an autonomous abort was conducted. More reliability was achieved through reducing the number of failure modes, by minimizing the number of separate subsystems. For example, the Falcon 1 first stage Thrust Vector Control (TVC) system made use of the pressurized fuel, RP-1, through a line tapped off the high-pressure RP side of the pump to power the TVC, a design that not only eliminated the separate hydraulic system but also eliminated the failure mode associated with running out of pressurized fluid.

All the Falcon 1's design decisions had been subject to exhaustive testing, ranging from component level qualification and workmanship testing, structures load and proof testing, to flight system and propulsion subsystem level testing. Adding another layer of ruggedization to their vehicle, SpaceX tested beyond the margins of environmental extremes and conducted stage and fairing separation tests for off-nominal events, such as geometrical misalignment, anomalous pyrotechnic timing, and sequencing.

Once a payload has been accepted for launch, the process of installing the payload features a general sequence of events. As you can imagine, all these procedural checks represented a steep administrative learning curve for SpaceX during the Falcon 1 years, so it is instructive to follow the steps in the process. First, each payload must interface with the launch vehicle by means of a Payload Attach Fitting (PAF), which can be modified to accommodate customer needs. Next, before the vehicle is shipped to the launch site, a mechanical fit check that includes electrical connector locations must be configured with the spacecraft. This check is conducted by SpaceX personnel during the payload integration process. The payload must also be checked for electrical design interface for ground and flight operations. Customers must also consider the environment their payload will be exposed to, from the time it leaves the customer's location until it is released in orbit. Much of this information is contained in the SpaceX Interface Control Document (ICD), which defines the various environments. For example, the first environment to consider is transportation, which is encountered by the payload during its journey from the payload processing hangar to the launch pad and may be accomplished by wheeled vehicle (and, in the Falcon 1 years, by ocean vessel). Since SpaceX does not control ambient temperature, humidity, and cleanliness during this journey, it is up to the customer to ensure their payload transportation containers are designed to protect the payload until it is removed from the container in the environmentally controlled payload processing facility. The next stage in this rather exhaustive process is looking after the payload while it is in the clean room before being stowed away within the fairing. The clean room is an air-conditioned facility maintained at 21°C ± 5.5°C and a humidity level between 30 and 60 percent. The payload remains in an air-conditioned environment after being encapsulated, thanks to air being provided via a flexible duct system to a fairing port configured to direct the air into the fairing. From this stage, except for a short break to move the payload from the clean room to the vehicle, the payload remains in an air-conditioned environment until launch.

Launch is the toughest environment the payload is subjected to during its journey from the customer to orbit. In fact, the launch loads the payload has to endure in the first ten minutes of the climb to orbit drive much of its structural design and mass. SpaceX is acutely aware of the stresses placed on a payload during launch and has gone to great lengths to reduce the Falcon family's launch loads. However, although SpaceX decided to forego the use of solid rocket boosters (a culprit in imposing high launch loads), there is no escaping the shock loads that occur during every flight. First, there is the shock of the hold-down release of the Falcon vehicles at lift-off, followed shortly thereafter by stage separation. While these shocks are negligible, the shock of fairing separation can be punishing due to the distance and number of joints over which the shocks travel and dissipate. Less punishing is the radiofrequency (RF) environment, but customers are still required to ensure that components sensitive to RF are compatible with the launch pad environment, which is characterized by frequencies governing command and destruct, tracking transponder, vehicle launch telemetry, GPS, UHF, C-band, and S-band. Other considerations include spacecraft fueling, which SpaceX accommodates as a non-standard service, and electrical power supply, which requires the customer to provide the necessary cables to interface the payload with the payload processing room. There is also the issue of monitoring the payload once it is in orbit. During test and launch operations, SpaceX provides one console for the customer in the SpaceX mission control center (Figure 4.1), and stations for up to five other payload support personnel during launch operations, either in the payload processing area or in other facilities.

Once SpaceX is happy with all the information and with the payload specification provided by the customer, and once the customer (the Payload Provider in SpaceX parlance) is happy with the services provided by SpaceX, the customer is assigned a Mission Manager, who serves as a point of contact from contract award through launch. It is the Mission Manager who assesses the launch vehicle capabilities against payload requirements, conducts mission design reviews and attends to the myriad teleconferences and integration meetings required before launching a payload into orbit (see Table 4.1). It is also the Mission Manager's job to coordinate other administrative aspects such as range and range safety integration and mission-required licensing. Once the payload arrives at the launch site, the physical accommodation for the spacecraft is turned over to the Payload Integration Manager, although the Mission Manager continues to manage the customer interface at the launch site.

Figure 4.1. Yes, SpaceX has its own mission control center, located in Hawthorne, California. Credit NASA.

Table 4.1: Generic Launch Integration Process

Timeline	Activity
T – 8 months	• Estimated payload mass, volume, mission, operations, and interface requirements.
	• Safety information; design information such as battery, ordnance, propellants, and operations.
	• Mission analysis summary provided to customer within 30 days of contract.
T – 6 months	• Final payload design, including mass, volume/structural characteristics, mission, operations, and interface requirements.
	• Payload to provide test verified structural dynamic model.
T – 4 months	Payload readiness review for Range Safety – includes launch site operations plan and hazard analyses.
T – 3 months	• Review of payload test data verifying compatibility with Falcon 1 environments.
	• Coupled payload and Falcon 1 loads analysis completed.
	• Confirm payload interfaces as built are compatible with Falcon 1 mission safety approval.
T – 4–6 weeks	**System Readiness Review (SRR)**
	• Pre-shipment reviews.
	• Launch site verified. Range, regulatory agencies, launch vehicle, payload, people, and paper in place and ready to begin launch campaign.
T – 2 weeks	Payload arrival at launch location.
T – 8–9 days	Payload mating to Launch Vehicle and fairing encapsulation.
T – 7 days	**Flight Readiness Review (FRR)**
	• Review of launch vehicle and payload checkouts in hangar.
	• Confirmation of readiness to proceed with Vehicle rollout.
T – 1 day	**Launch Readiness Review (LRR)**
Launch	
T + 4 hours	Post-Launch Reports – Quick look
T + 4 weeks	Post-Launch Report – Final Report

Once the customer's payload has made it to the launch site, SpaceX makes its pre-launch operations as simple and streamlined as possible, a sequence of events that begins 18 days prior to launch. Before a Flight Readiness Review (FRR) can be completed, payload attachment and fairing encapsulation must be finished. The process normally takes less than 24 hours and begins by integrating the payload on the adapter in the vertical configuration, followed closely by fairing encapsulation (Figure 4.2a, b, c). Once fully encapsulated, the system is rotated horizontally and integrated to the second stage. When this has been completed, post-mate checkouts can be conducted, followed by the FRR. With the FRR completed, the vehicle is then rolled out to the pad. Six days prior to launch, the integrated payload and launch vehicle are positioned vertically using the Launch Vehicle Transporter. Final system close-out, fueling and testing is then completed. The Launch Readiness Review (LRR) is held 24 hours prior to launch. Once the launch approval is given, the 24-hour countdown begins.

Figure 4.2a. The Transiting Exoplanet Survey Satellite (TESS) being prepared for securing within the Falcon 9's 5.4-meter diameter, 13-meter-long payload fairing, which will protect the spacecraft while sitting on the launch pad and shield it from aerodynamic forces on the way through the atmosphere. Credit NASA.

Figure 4.2b. Engineers prepare to close the fairing. Credit NASA

Figure 4.2c. The TESS payload is readied for transport to the pad. Credit NASA

In keeping with the company's mode of operation, the development of Falcon 1 was rapid, with fabrication of a prototype vehicle beginning in early 2003. Within the year, SpaceX unveiled the vehicle – after driving it cross-country on its custom-built transport trailer – in Washington, D.C., in December 2003, parking it on the street in front of the Federal Aviation Administration (FAA) building. Musk used the occasion to pronounce that his company was planning a sequel to the Falcon 1 by building a more powerful 3.7-meter diameter Falcon 5, to be powered by five Merlin engines. Initial pricing for the Falcon 1 was set at $6 million, while Falcon 5 – designed to haul 4.5 tonnes to LEO – was listed at $12 million. In October 2004, the SpaceX company made plans to launch its first Falcon 1 at their SLC 3W launch pad at Vandenberg Air Force Base (VAFB). In the months that followed, SpaceX built up a waiting list of missions, including one to launch a U.S. Navy microsatellite called TacSat-1, and another to orbit a test payload for Bigelow Aerospace. These contracts were impressive considering that SpaceX had yet to launch a single vehicle. To do so, they had to complete development of the Merlin engine and then the Falcon 1 vehicle itself had to be verified. This was achieved with a series of structural tests in March 2005. With these milestones reached, SpaceX was ready to launch TacSat-1, but military bureaucracy got in the way because the Air Force did not want SpaceX to launch until a Titan 4 had flown from nearby SLC 4E. After repeated delays pushed the Titan launch back, an exasperated Musk decided to fly the first Falcon 1 from Kwajalein instead and, in June 2005, SpaceX ferried the Falcon launch equipment to Omelek in the Marshall Islands, followed by the first Falcon 1 vehicle a month later.

The first launch attempt at Omelek on November 25, 2005, did not go well, the launch effort being scrubbed after a ground-supply LOX vent valve allowed the LOX supply to boil off. A second attempt was made on December 19, 2005, but this was delayed by high winds. Worse was to come when the first stage fuel tank buckled during fuel draining thanks to the fuel pressurization system having suffered a controller failure. The damaged first stage had to be shipped to Los Angeles for repair and was replaced with the first stage from the second Falcon 1. Less than two months later, SpaceX tried again. On February 9, 2006, the company completed a hot-fire test at the Omelek pad with the new first stage. Unfortunately, a second stage propellant leak was discovered during the testing process, which halted the February launch attempt. Again, the company shipped the troublesome stage back to Los Angeles, this time replacing it with the second stage of the second Falcon 1, reuniting it with its first stage. On March 18 and 23, 2006, SpaceX ramped up for a fourth launch attempt by performing hot-fire tests using the reconfigured vehicle.

Having successfully performed the hot-fire tests, everything seemed to be on track for a successful inaugural launch on March 24, 2006. After a 22:30 GMT liftoff, the Falcon rose from its pad and ascended in what appeared to be a clean, stable climb. The launch proceeded according to the script until about 25 seconds into flight, when the fire just above the engine cut into the first stage helium pneumatic system, causing an engine shut-down at T+34 seconds. Following the shutdown, the Falcon 1 rolled and fell into the ocean. Since the Falcon 1 was equipped with an engine cut-off range safety system rather than destruct charges, the vehicle fell mostly intact onto a reef not far from the launch site. SpaceX immediately went to work troubleshooting the problem. A week later, in an NPR interview, company Vice President (VP) Gwynne Shotwell informed the news media that the leak had been caused by a procedural error rather than a hardware failure of the Falcon 1. Four months after the failed launch, SpaceX reported the findings of a Defense Advanced Research Projects Agency (DARPA) *Falcon Return to Flight Board*, which revealed that a kerosene fuel leak had begun 400 seconds before liftoff when the propellant pre-valves were opened. When the Merlin main engine had started at liftoff, the leaking fuel had ignited.

As SpaceX prepared for its next Falcon 1 launch attempt, it announced revised design information for its launch vehicle, providing details on its website of a new Merlin 1C-powered Falcon 1e rocket that would be 5.53 meters taller and 11.36 tonnes heavier than the original Falcon rocket. The Falcon 1e, which was expected to enter service after 2009, would haul 25–30 percent more payload than the Falcon 1.

The next attempt to launch the Falcon 1 was scheduled for early in 2007. After being erected at Omelek Island in mid-January 2007, a late-January hot-fire test was postponed when the vehicle's second stage engine failed a slew test during the countdown. On March 15, 2007, Space X performed a successful static test ignition of the Falcon 1 first stage Merlin engine. This was followed by a scrubbed launch attempt on March 19, 2007. Two days later, the Falcon 1 failed to reach orbit after flight control was lost two minutes into the vehicle's second stage burn. While not completely successful[2], the flight did achieve several milestones, including passing through Max-Q, completing a first stage burn, stage separation, second stage ignition, and jettisoning of the payload fairing. The launch also demonstrated the operational responsiveness of SpaceX thanks to a launch abort that stopped the main engine start sequence. The abort, caused by a low chamber pressure reading resulting from lower than planned kerosene fuel temperatures, required SpaceX crews to drain and reload some of the first stage fuel. The glitch allowed the SpaceX crew to demonstrate just how quickly they could reload propellant, with the whole exercise taking less than an hour before restarting the

---

[2] At launch, the Falcon 1 weighed 27.526 tonnes. First stage burnout occurred 168 seconds after liftoff at an altitude of 75 km and a velocity of 2.6 kilometers per second. The second stage Kestrel engine ignited five seconds after first stage cutoff, beginning a planned burn of 415 seconds intended to insert the stage into an initial 330 x 685 km orbit about 585 seconds after liftoff. Ultimately, the second stage achieved a suborbital velocity of about 5.1 kilometers per second, reaching a maximum altitude of 289 km.

count. The reason for the partial failure was captured on the on-board video broad-cast, which showed the second stage engine bell brushing against the side of the inter-stage at stage separation. The video also revealed an oscillatory motion developing during the last minute of controlled flight, just before roll control and telemetry was lost. Six days later, Musk explained that LOX sloshing had caused the oscillation; the LOX slosh frequency had coupled with the thrust vector control system in a way that amplified the oscillation until flight control was lost.

In April 2008, SpaceX revealed details of a beefed-up Merlin 1C capable of producing more than 56 tonnes of sea-level thrust, a capability that translated into a 1-tonne LEO payload for the Falcon 1e (by comparison, the Falcon 1's LEO payload was advertised as 0.42 tonnes). Later that year, on August 3, the third Falcon 1 rocket failed shortly after lifting off from Omelek Island, with the cause of the failure being attributed to residual thrust produced by the Merlin 1C first stage engine which caused the stage to recontact the second stage immediately after stage separation. Lost with the Falcon 1 were the USAF's Trailblazer satellite, NASA's Nanosail-D solar sail experiment, and NASA's PreSat experiment. Echoing the second attempt, the third launch effort also suffered a countdown abort, this one 34 minutes before launch. Demonstrating their efficiency once again, SpaceX crews recycled the count in 23 minutes. Less than two months later, on September 28, SpaceX was ready to go again. This time the launch went as advertised and the company's fourth Falcon 1, carrying a 165 kg payload mass simulator, reached a 330 x 650 km orbit. There were a few minor deviations from the plan, the first being a slightly lower orbit than the planned 330 x 685 kilometers. Another minor departure from the flight profile was a slightly premature shutdown of the second stage, though this was compensated for by the Kestrel second stage engine performing a test of its restart capability in space. But these were minor issues; the major objective had been achieved and the Falcon 1 booster had redeemed itself with an electrifying launch that put an exclamation point on six years of hard work and occasional setbacks for SpaceX. The mission logo for the launch, known as Flight 4 in SpaceX parlance, included two four-leaf clovers, symbolizing the end of the rocket's string of bad luck.

After flights plagued by delays and last-second aborts, the successful launch of the Falcon 1 marked a major milestone and accomplished a feat that only a handful of countries had achieved. Priced at just $7.9 million per flight, the Falcon 1 was three times less expensive than its U.S. competitors in the commercial launch market, as well as its Russian, Chinese and Indian rival launch vehicles. After the inevitable setbacks and delays that are a fact of life when trying to develop a new rocket, SpaceX had demonstrated it had the patience and the knowledge to adapt and innovate after every launch failure and, in so doing, deliver on its promise of putting the company on the map of commercial spaceflight. Having demonstrated that they could get into orbit, the company's next objective was to show they could repeat the feat.

On July 14, 2009, SpaceX launched its fifth Falcon 1, boosting the RazakSAT, a Malaysian government Earth observation imaging satellite, into orbit. Following a 2 minute 40 second burn, the first stage fell away and the second stage Kestrel engine ignited, completing its first burn about 9 minutes 40 seconds after liftoff and boosting the stage and 180-kilogram payload toward a 330 x 685-kilometer parking orbit. The launch marked the final original Falcon 1 on the SpaceX manifest and the first launch of a live satellite. The smooth RazakSAT flight had underscored the company's position in the tough world of commercial spaceflight. In just seven years of existence, SpaceX had positioned itself as a force to be reckoned with, an achievement that was only possible because it was a private company. Because the company had not been encumbered by government red tape, Musk had been able to develop SpaceX as he wanted – as long as the money was there, which it was thanks in part to NASA and its COTS Program.

After so much work and money developing and flight-proving the Falcon 1, the expectation was that SpaceX would begin to exploit its new vehicle's operational status and aggressively market and sell its new launch service. But that was the end of the Falcon 1. For SpaceX to spend six years and millions of dollars to develop its launch system, only to abandon it just as success and profit was at hand, is difficult to understand. Was the development of the Falcon 1 as a commercial launch system never intended in the first place? In an interview with NASASpaceflight.com, when SpaceX was asked if there was still a future role for Falcon 1/1e, the company's communications director Kirstin Brost Grantham said that their plans were for small payloads to be served by flights on the Falcon 9, utilizing excess capacity. In other words, SpaceX did not see a future for the Falcon 1e. It was the logical next step, given that the Falcon 9 can launch small satellites perfectly well as secondary payloads.

So, the Falcon 1 has been retired. Its legacy? Well, it depends on your viewpoint. From the perspective of launch successes, the Falcon 1 had one of the all-time *worst* orbital launch vehicle records in history; it failed three times in five attempts and managed to send only one satellite into orbit – and even that satellite failed to function properly. That is a terrible launch record and one that is only rivalled by India's GSLV[3] and Iran's Safir (which failed three times out of four) among modern rockets that have flown more than twice. But from a legacy perspective, the Falcon 1 represents solid engineering and the results of that were transferred to the development of the Falcon 9, which has proven very, *very* successful.

---

[3]The Geosynchronous Satellite Launch Vehicle (GSLV) is an expendable launch system operated by the Indian Space Research Organization (ISRO) and was developed to launch satellites into geostationary orbit. Since its first launch in 2001 there have been seven GSLV launches: two successful, one partially successful and four failures.

5

Falcon 9 and Falcon Heavy

Life after Shuttle

"The Space Shuttle has changed the way we view the world and it's changed the way we view our universe. There's a lot of emotion today, but one thing's indisputable. America's not going to stop exploring."

STS-135 Space Shuttle Commander Chris Ferguson.

Figure 5.0. A SpaceX Falcon 9 rocket and Dragon spacecraft at Launch Complex 39A ready for the tenth SpaceX Commercial Resupply Services (CRS) mission to the International Space Station (ISS). The mission set a milestone, as the first launch from Launch Complex 39A since the Shuttle retired in 2011. The launch also marked a turning point for the Kennedy Space Center's (KSC) transition to a multi-user spaceport geared for commercial missions, as well as those conducted in partnership with NASA. Credit NASA.

© Springer Nature Switzerland AG 2022
E. Seedhouse, *SpaceX*, Springer Praxis Books,
https://doi.org/10.1007/978-3-030-99181-4_5

It was a hot summer day on Florida's Space Coast as nearly a million spectators gathered along the beaches and causeways to watch history in the making. A last-minute glitch held the clock at T-31 seconds, just as launch looked imminent, but the issue was quickly resolved and the clock began counting down the final seconds. With less than a minute remaining in the launch window, the three main engines of Shuttle *Atlantis* roared to life and the twin Solid Rocket Boosters (SRBs) thundered. For the thirty-third and final time, *Atlantis* rose majestically from the launch pad on a plume of fire (Figure 5.1) and parted the clouds on its way to the International Space Station (ISS) and to its place in history. The 11:29 a.m. EDT liftoff on July 8, 2011, marked the last time a Shuttle would climb from the Kennedy Space Center (KSC) launch complex.

Less than 13 days later, on July 21, *Atlantis* made an elegant sweeping turn, lined up with the runway and landed for the final time, in the half-light before dawn at KSC. Wheel stop came at 5:58 a.m. EDT, after a flight of 12 days, 18 hours, 28 minutes and 55 seconds. And that was that. After 135 flights in 30 years and having travelled more than 800 million kilometers in orbit, the Shuttle era was

Figure 5.1. The Shuttle *Atlantis* launches from pad 39A on July 8, 2011, on the final Shuttle mission, STS-135.

history. *Atlantis* alone made 33 flights, carried 191 astronauts, spent 307 days in orbit, circled Earth 4,848 times and clocked more than 200 million kilometers. With *Atlantis* on the ground, 2,300 Shuttle workers received layoff notices and more than 8,000 people who worked for NASA or its contractors in the Shuttle program lost their jobs. It was a quiet ending to a program that was supposed to have made spaceflight affordable, safe and routine. Instead, it proved risky and expensive, with each Shuttle flight costing more than one billion dollars. The final landing of *Atlantis* left NASA with no spacecraft to send its astronauts to orbit. As a stopgap measure, the agency signed a $763 million contract for 12 Russian rocket rides (that's more than $63 million per flight) from 2014 through 2016. By that time, the space agency hoped at least one of the four private companies it was seeding with cash would demonstrate a crew-ready launcher. But by 2017 it was clear that delays to both the SpaceX Crew Dragon and the Boeing Starliner CST-100 meant NASA would have to pay the Russians for more tickets (Figure 5.2). In February 2017, the agency paid Roscosmos $373.5 million for five seats, or $74.7 million per seat, and prayed fervently that either the Dragon or the CST-100 would be ready by 2019.

Figure 5.2. Between 2011 and 2020, NASA, ESA, the CSA and JAXA relied on the Russian Soyuz to ferry its astronauts to the ISS. After the Shuttle retired, NASA bought flights for its astronauts on Soyuz at a cost of between $63 and $90 million per seat. In May 2020, NASA announced it had bought its final flight on the Soyuz for $90 million ($90,252,905.69 to be exact). In the decade since the Shuttle's retirement, the Soyuz had carried 39 American astronauts on 36 missions. Credit NASA.

Figure 5.3. A Falcon 9 rocket and Crew Dragon *Resilience*, ready for the Crew-1 mission inside the SpaceX Hangar at Kennedy Space Center on November 9, 2020. Credit NASA.

Returning to the retirement of the Shuttle, attention was focused on the Falcon 9 as one way of launching crews to orbit. The Falcon 9 (Figure 5.3) is a two-stage rocket powered by liquid oxygen (LOX) and rocket-grade kerosene (RP-1). In common with the Falcon 1, it was designed from the ground up by SpaceX for the cost-efficient transportation of satellites to Low Earth Orbit (LEO) and Geosynchronous Transfer Orbit (GTO), and for sending the Dragon and Crew Dragon, carrying cargo and/or astronauts, to orbiting destinations such as the ISS. It is a 'Made in America' rocket, with all structures, engines, avionics, *and* ground systems designed, manufactured and tested in the United States by SpaceX. When used to carry crew, the Falcon 9 is 48.1 meters tall with a Dragon perched on top. Developed from a blank sheet to first launch in just four-and-a-half years (November 2005 to June 2010), the Falcon 9 generates one million pounds of thrust in a vacuum (see Table 5.1). The cost? $62 million new and about $50 million used if you want to save some money (each flight costs about $1 million to insure, incidentally). The vehicle features cutting-edge technology and a simple two-stage design to limit separation events. With nine regeneratively-cooled engines on the first stage (hence the number '9' in the designation), the Falcon 9 can still safely complete its mission in the event of an engine failure. The tank walls are made from an aluminum-lithium alloy that SpaceX manufactures using friction-stir welding,

Table 5.1: Falcon 9 Specifications

Height	FT: 70 m
	v1.1: 68.4 m
	v1.0: 54.9 m
Diameter	3.7 m
Mass	FT: 549,054 kg
	v1.1: 505,846 kg
	v1.0: 333,400 kg
Payload to LEO (28.5°)	FT: 22,800 kg expended
	v1.1: 13,150 kg
	v1.0: 10,450 kg
Payload to GTO (27°)	FT: 8,300 kg expended
	v1.1: 4,850 kg
	v1.0: 4,540 kg
Payload to Mars	FT: 4,020 kg
First flight	FT Block 5: May 11, 2018 (Bangabandhu-1)
	FT: December 22, 2015 (OG2 Flight 2)
	v1.1: September 29, 2013 (CASSIOPE)
	v1.0: June 4, 2010 (Dragon COTS Demo 1)
First stage	
Engines	Block 5: 9 Merlin 1D+ (max thrust)
	FT: 9 Merlin 1D+
	v1.1: 9 Merlin 1D
	v1.0: 9 Merlin 1C
Thrust	FT: 6,806 kN (1,530,000 lb$_f$)
	v1.1: 5,885 kN (1,323,000 lb$_f$)
	v1.0: 4,940 kN (1,110,000 lb$_f$)
Specific impulse	v1.1 Sea level: 282 seconds (Vacuum: 311 seconds)
	v1.0 Sea level: 275 seconds (Vacuum: 304 seconds)
Burn time	FT: 162 seconds
	v1.1: 180 seconds
	v1.0: 170 seconds
Fuel	LOX / RP-1
Second stage	
Engines	FT: 1 Merlin 1D Vacuum+
	v1.1: 1 Merlin 1D Vacuum
	v1.0: 1 Merlin 1C Vacuum
Thrust	FT: 934 kN
	v1.1: 801 kN
	v1.0: 617 kN
Specific impulse	FT: 348 seconds
	v1.1: 340 seconds
	v1.0: 342 seconds
Burn time	FT: 397 seconds
	v1.1: 375 seconds
	v1.0: 345 seconds
Fuel	LOX / RP-1

which is the strongest and most reliable welding technique available. Connecting the lower and upper stages is the interstage, a composite structure with an aluminum honeycomb core and carbon fiber face sheets. The separation system is pneumatic, a system proven on its predecessor, the Falcon 1.

The Falcon 9 second stage tank is a shorter version of the first stage tank and uses many of the same tooling, material and manufacturing techniques, a policy that results in significant cost savings in vehicle production. Powering the upper stage is a single Merlin engine, capable of restart thanks to dual redundant pyrophoric igniters using triethyl aluminum-triethylborane (TEA-TEB). The vehicle is a reliable system, in part because it only has only two stages which limits problems associated with separation events. This reliability is enhanced by the advanced avionics package, which features a hold-before-release system. This is a capability required by commercial airplanes, but is not implemented on many launch vehicles. This is how it works. After the first-stage engine ignites, the Falcon 9 (Table 5.2) is held down and not released for flight until all propulsion and vehicle systems are operating nominally. If any issues are detected, an automatic safe shut-down occurs and propellant is unloaded.

Table 5.2: Falcon 9 Performance

Version	Falcon 9 v1.0 (retired)	Falcon 9 v1.1 (retired)	Falcon 9 Full Thrust Blocks 3/4 (retired)	Falcon 9 Block 5 (active)
Stage 1	9 × Merlin 1C	9 × Merlin 1D	9 × Merlin 1D (upgraded)	9 × Merlin 1D (upgraded)
Stage 2	1 × Merlin 1C Vacuum	1 × Merlin 1D Vacuum	1 × Merlin 1D Vacuum FT	1 × Merlin 1D Vacuum FT
Height (m)	53	68.4	70	70
Diameter (m)	3.66	3.66	3.66	3.66
Initial thrust (kN)	3,807	5,885	6,804	7,600
Takeoff mass (tonnes)	318	506	549	549
Fairing diameter (m)	N/A	5.2	5.2	5.2
Payload to LEO (kg)	8,500–9,000	13,150	22,800	22,800
Payload to GTO (kg)	3,400	4,850	8,300 (expendable) 5,300 (reusable)	8,300 (expendable) 5,500 (reusable)

Capable of lifting payloads of 10,450 kilograms to LEO, and 4,450 kilograms to GTO, the Falcon 9 launched on its maiden flight – after several delays – from Cape Canaveral Air Force Station (CCAFS) on June 4, 2010, at 2:45 pm EDT (see Table 5.3). Liftoff came 3 hours and 45 minutes into a four-hour launch window, because of tests conducted on the rocket's self-destruct system and a last-second abort caused by a higher-than-expected pressure reading in one of the engines.

Table 5.3: Falcon 9, Flight #1 Launch Timeline

T+ 00:00:06	Liftoff.
T+ 00:01:13	Falcon 9 approached maximum dynamic pressure (Max Q).
T+ 00:01:56	Plume behind the vehicle expands as the atmosphere thins.
T+ 00:03:06	Stages separated.
T+ 00:03:34	Second Stage Ignition.
T+ 00:08:50	Second Stage Engine Shut-Down.
T+ 00:09:04	Falcon 9 achieved Earth orbit, delivering a dummy payload, a structural test article representing the company's planned Dragon space station cargo module.

SpaceX engineers resolved the issue and recycled the countdown to the T-minus 15-minute mark, after concluding that the engine was in good shape.

Telemetry from the Falcon 9 confirmed the vehicle had achieved an orbit of almost exactly 250 km. If you consider that two-thirds of the rockets introduced in the past 20 years have had an unsuccessful first flight, the maiden Falcon 9 flight could not be considered as anything other than 100 percent successful. Elon Musk and his team of engineers were not the only ones breathing a sigh of relief as the Falcon 9 soared above their heads; NASA, which had signed contracts with SpaceX totaling $1.6 billion for ISS deliveries, also began to feel a little more comfortable. The successful launch not only augured well for the U.S. shift in national space priorities (turning launches to LEO over to the private sector while NASA focused on deep space exploration) but also vindicated Musk's approach by demonstrating that a new company could make a difference. Talking to the media following the launch, Musk acknowledged NASA's help, but also pointed out that the first Falcon 9 heralded the dawn of new era of spaceflight, which would increasingly be marked by combined commercial and government endeavors and with commercial companies playing an increasingly significant role.

But although the flight was no doubt historic, it was difficult at the time to judge whether it was a game changer. After all, SpaceX could not have succeeded without significant government money and the promise of a healthy supply contract, and they had received plenty of technical support from NASA. Having said that, for many, the first Falcon 9 flight represented a step away from the duplicative and appallingly wasteful government way of doing business and provided a stimulus for other companies such as Blue Origin (with their New Glenn launch vehicle) to follow in the pioneering path of SpaceX. It was also worth remembering that other companies had followed a similar route to SpaceX and failed. For example, Orbital Sciences Corporation (Orbital) was considered the latest and greatest spaceflight company in the early 1990s. They pronounced they would reduce the cost of going into space but ultimately were absorbed into the system and became just another overpriced aerospace company. With regard to cost, it is instructive to compare vehicles in terms of dollars per kilogram lifted. Compared to the Shuttle, SpaceX had a good case for saying they could launch for one-tenth of the cost, but compared to the Atlas V551 or the Delta IV Heavy, that ratio was probably closer to one-fifth. Unlike the Delta and Atlas however, which only launch once or twice per year, the SpaceX policy of mass production and paralleling of systems over the years of development has meant that economies of scale have crept in, allowing Musk to advertise a Falcon 9 launch (using a reused first stage) for just $50 million.

"When Dragon returns, whether on this mission or a future one, it will herald the dawn of an incredibly exciting new era in space travel. This will be the first new American human-capable spacecraft to travel to orbit and back since the Space Shuttle took flight three decades ago. The success of the NASA COTS/CRS program shows that it is possible to return to the fast pace of progress that took place during the Apollo era but using only a tiny fraction of the resources. If COTS/CRS continues to achieve the milestones that many considered impossible, thanks in large part to the skill of the program management team at NASA, it should be recognized as one of the most effective public-private partnerships in history."

Elon Musk, SpaceX CEO & CTO, speaking shortly before the second Falcon 9 flight.

On December 8, 2010, SpaceX became the first commercial company in history to return a spacecraft from Earth orbit (see Table 5.4). The second Falcon 9 flight – the first under the COTS Program – began from Launch Complex 40 at the CCAFS in Florida and, having followed a nominal flight profile that included a 9.5-minute ascent and two Earth orbits, ended with the Dragon reentering the Earth's atmosphere a few hours later, landing less than one mile from the center of the landing zone in the Pacific Ocean.

Before the second Falcon 9 flight, the act of recovering a spacecraft reentering from Earth orbit was a feat that had previously been achieved by only six nations or government agencies: the United States, Russia, China, Japan, India, and the European Space Agency (ESA). In addition to achieving the Dragon's first-ever on-orbit performance, eight free-flying payloads were successfully deployed, including a U.S. Army nanosatellite that was the first Army-built satellite to fly in 50 years. As with its maiden flight, the Falcon 9 rocket performed nominally during ascent and staging. After separation of the Dragon, the second stage Merlin engine restarted, carrying the second stage to an altitude of 11,000 km and confirming SpaceX's capability of achieving GTO missions. As the Dragon passed over Hawaii, SpaceX received video sent from the spacecraft on orbit. Then, in preparation for re-entry, the Draco thrusters, each capable of producing about 90 pounds of thrust, began the six-minute de-orbit burn. All thrusters performed nominally, although the Dragon could still have returned to Earth even if two quads had been lost. During re-entry, the Dragon's Phenolic Impregnated Carbon

Table 5.4: Falcon 9, Flight #2 Launch Timeline

Countdown	
T-02:35:00	Chief Engineer polled stations. Countdown master auto-sequence proceeded with LOX load, RP-1 fuel load, and vehicle release.
T-01:40:00	Master auto sequence proceeded into lowering the strongback.
T-00:60:00	Master auto sequence proceeded with stage 2 fuel bleed, stage 2 thrust vector control bleed. Verification of all sub-auto sequences in the countdown master auto sequence, except for terminal count.
T-00:13:00	SpaceX Launch Director polled readiness for launch.
T-00:11:00	Logical hold point.
Terminal Count (begins at T-10 minutes)	
T-00:09:43	Pre-valves opened to the nine first stage engines to begin chilling Merlin engine pumps.
T-00:0-6:17	Command flight computer entered alignment state.
T-00:05:00	Loading of GN2 into ACS bottle on stage 2 ceased.
T-00:04:46	Internal power on stage 1 and stage 2 transferred.
T-00:03:11	Arming flight termination system began.
T-00:03:02	Terminated LOX propellant topping and cycled fuel trim valves.
T-00:03:00	Verified movement on stage 2 thrust vector control actuators.
T-00:02:30	SpaceX Launch Director verified – GO.
T-00:02:00	Range Control Officer (Air Force) verified range is – GO.
T-00:01:35	Terminated helium loading.
T-00:01:00	Commanded flight computer state to start-up.
T-00:01:00	Turned on pad deck and Niagara water.
T-00:00:50	Flight computer commanded thrust vector control actuator checks on stage 1.
T-00:00:40	Pressurized S1 and S2 propellant tanks.
T-00:00:03	Engine controller commanded engine ignition sequence to start.
T-00:00:00 Liftoff	
T+0:02:58	1st Stage Shut Down (Main Engine Cut Off).
T+0:03:02	1st Stage Separated.
T+0:03:09	2nd Stage Engine Started.
T+0:09:00	2nd Stage Engine Cut-off.
T+0:09:35	Dragon Separated from Falcon 9 and initialized propulsion.
T+0:13	On-Orbit Operations.
T+2:32	Deorbit Burn Began.
T+2:38	Deorbit Burn Ended.
T+2:58	Re-entry Phase Began (Entry Interface).
T+3:09	Drogue Chute Deployed.
T+3:10	Main Chute Deployed.
T+3:19	Water Landing.

Ablator (PICA-X)[1] heat shield protected the spacecraft from temperatures exceeding 1,650°C. At 3,000 meters, Dragon's three main parachutes deployed, slowing the spacecraft's descent and demonstrating the safe landing capability required for

---

[1] SpaceX had worked with NASA to develop the 3.6-meter PICA-X, a variant of NASA's PICA heat shield. Designed, developed and qualified in less than four years at a small fraction of the cost NASA had budgeted for the effort, it is the most advanced heat shield ever to fly and was designed to be used hundreds of times with only minor degradation.

manned flights. In common with the level of redundancy of many of the vehicle's systems, Dragon could have lost one of its main parachutes and the two remaining chutes would still have ensured a safe landing.

Given the success of the second Falcon 9 flight, it was no surprise when SpaceX proposed the possibility of merging two scheduled COTS flight demonstrations of the Falcon 9/Dragon combination. It was planned that the combined flight would precede routine resupply runs to the ISS under a separate $1.6 billion fixed-price contract with NASA. To facilitate the previously unplanned ground tests needed to support the combined demonstration flight, the agency boosted its investment in SpaceX by $128 million in 2011. The combined flight made sense, because at the time SpaceX was more than two years behind in completing its COTS demonstration flights. In the original plan the company's second COTS demo had been a five-day mission, during which Dragon would have approached to within 10 kilometers of the ISS and used its radio cross-link to allow the station's crew to receive telemetry from the capsule and send commands. In the third and final COTS demo, Dragon would have berthed with the ISS for the first time.

Before a decision to merge the two COTS demo missions could be approved by NASA, however, SpaceX needed to conduct a full-up thermal vacuum test and electromagnetic interference test of Dragon, in addition to completing a closed-loop demonstration of the capsule's proximity operations sensors. They also needed to upgrade its production operation in Hawthorne, California, add engine test stands at its facility in McGregor, Texas, *and* make improvements to its launch pad at the Cape. Fortunately, SpaceX and fellow COTS provider Orbital had earned a combined $80 million in milestone payments since the beginning of the 2010 fiscal year, money that came in addition to the payouts negotiated in their original COTS agreements. Also, while both companies were running behind in meeting milestone objectives, they had received the bulk of the funding anticipated under their original COTS contracts, which totaled $448 million combined.

The extra tests were necessary due to the length of the proposed mission. Whereas Dragon had spent just a few hours in orbit on its maiden flight, the merged mission would extend its time on orbit to several days. This extra time in LEO would expose Dragon to several long cold-soak and long hot cycles, a thermal stress environment that had not been tested during the first Dragon flight. In addition to the thermal tests, SpaceX had to fine-tune the rendezvous and proximity operations software required for Dragon's berthing with the ISS. These extra tasks required an improvement in the throughput at the company's facilities, work that NASA, as an investor, was only too happy to support by providing SpaceX with another $10 million COTS payout for developing a plan to improve its production, test and launch facilities. Meanwhile, Orbital, a SpaceX competitor in the private space race which was developing the Taurus 2 rocket and Cygnus spacecraft under its $170 million original COTS agreement, received an extra $40 million in new milestone payments to help the company prepare for an additional test flight of the Taurus 2, which would carry a dummy Cygnus capsule aloft. Under Orbital's original COTS agreement, the company had been slated to conduct a single test flight of Taurus 2 that would deliver a cargo-laden Cygnus capsule to the ISS, but Orbital decided to bump the mission to the end of 2011 to make room for the additional flight demo.

With Space X and Orbital each planning to launch cargo to the ISS, 2011 was shaping up to be another notable year in the world of commercial spaceflight, as both companies were working hard towards realizing their ambition of ultimately ferrying astronauts to the ISS. Unfortunately, the year did not proceed according to plan for either company. First, SpaceX had to go to court over the nasty business of an industry consultant allegedly spreading rumors that the company's rockets were unsafe. In June 2011, SpaceX filed with the Fairfax County circuit court in Virginia, alleging that Joseph Fragola, VP at tech consulting firm Valador, had tried to obtain a hefty deal worth as much as $1 million from SpaceX at the beginning of June. Fragola had claimed SpaceX needed an independent analysis of the Falcon 9 to bolster its reputation with NASA. Not surprisingly, SpaceX did not take kindly to Fragola's claims, especially when the company learned that the Valador VP had been contacting officials in the United States government to make negative remarks about SpaceX, thus creating the very perception that Fragola had claimed SpaceX needed his help to rectify. For example, one of the e-mails that Fragola wrote to Bryan O'Connor, a NASA official at the agency's headquarters, stated that he had heard a rumor that the second Falcon 9 flight had experienced a double engine failure in the first stage and that the stage had blown up after the first stage separated. SpaceX responded by saying that Fragola's rumor was false, explaining that two of the nine engines in the Falcon's first stage had shut down according to plan ten seconds before the other seven, and that there had been no engine failure. Because of Fragola's statements, SpaceX sued for $1 million for defamation from Valador.

The Valador case was not the first time SpaceX had been subject to industry bad-mouthing. Shortly before Fragola's statements, Loren Thompson, a paid consultant for big aerospace companies, had tried to sow doubt about the Falcon 9 and Dragon spacecraft to make it seem that NASA was betting the farm on an unproven company. Such a statement ignored the fact that NASA had diversified its portfolio of players, including Boeing and, of course, Orbital. It also sidestepped the fact that SpaceX had always publicly stated its support for competition. Thompson also made statements about how SpaceX had missed its schedule, while failing to acknowledge the innumerable slips of major government developments such as the short-lived and now defunct Constellation Program, or the fact that the Shuttle had been several years late to the pad. Another of Thompson's misplaced claims was that SpaceX was breaking its budget. Of course, this was also untrue since, as a commercial provider under COTS, SpaceX only received money when the company met its performance-based milestones. This was in contrast to Lockheed

Martin (one of Thompson's benefactors), the developer of the Orion capsule which had already cost upwards of $5 billion and was still many years and billions of dollars from completion.

Shortly after the Fragola legal case, Orbital announced that the inaugural flight of their Taurus 2 rocket would be delayed until December 2011, to allow them time to complete and certify the rocket propellant and pressurization facilities at the vehicle's Wallops Island launch site. The maiden flight would be followed by a second Taurus 2 flight two months later, a demonstration of its space station cargo vehicle in which the Cygnus capsule would approach the ISS. The delay stemmed from a June test failure of a Taurus 2 first stage AJ26 engine, which caught fire while being tested due to a fuel line breakage. While the company announced that neither the delay in verifying the propellant facility nor the engine failure would have a financial impact on their Taurus 2 program, the test failure was another setback for Orbital, which was under contract to make eight cargo delivery runs to the ISS starting in 2012. The company had not had a good start to 2011 following the failure, in March, of its smaller Taurus XL rocket whose fairing had malfunctioned for the second time. In both cases, the principal payloads had been NASA science satellites whose combined cost had been estimated at more than $600 million (the company later successfully placed the U.S. Defense Department's Operationally Responsive Space-1 satellite into LEO using a Minotaur rocket, a converted ICBM, which used a fairing that had taken into account the two Taurus XL failures and been redesigned).

The month after Orbital had announced its flight delay, SpaceX settled its lawsuit against Fragola. In August, SpaceX and Valador, Inc. said in a joint statement that both sides had agreed to terminate the lawsuit, although terms of the settlement were not disclosed.

The year ended without any flights by either SpaceX or Orbital. In December, Orbital decided to rebrand its new commercial rocket, changing the booster's name from the Taurus 2 to Antares and stating that the name change was to provide clear differentiation between Antares and the Taurus XL. The name *Antares*, which comes from the supergiant star located in the Scorpius constellation, was picked by Orbital because that is one of the brightest stars, and the company expressed the hope the Antares rocket would turn out to be one of the brightest stars in the space launch vehicle market. The name was also in keeping with the company's tradition of using Greek-derived celestial names for its launch vehicles. After renaming its rocket, Orbital announced it would fly the vehicle twice in the first half of 2012, before beginning a series of operational Cygnus resupply missions to the ISS later in the year. NASA had awarded Orbital a deal covering eight flights, valued at $1.9 billion, to deliver cargo to the station through 2015.

While Orbital was kept busy readying the Antares for the first of its two test flights, SpaceX continued its preparations for the third Falcon 9 flight. The second COTS flight would ferry Dragon to the ISS, where it would perform a series of rendezvous and approach maneuvers before delivering cargo to the orbiting outpost. If successful in its first-of-a-kind mission, SpaceX would collect the remaining payments on the $396 million contract it had with NASA and then enter into a $1.6 billion agreement for 11 more flights to the ISS.

The first launch attempt, on May 19, 2012 resulted in a countdown abort at T−00:00:00.5. Following the abort, SpaceX announced the next attempt would be scheduled three days later at 03:44 EDT, or on May 23, 2012 at 03:22 EDT. This second attempt was successful and three days later, after the gumdrop-shaped capsule had performed the requisite rendezvous and approach maneuvers, SpaceX became the first commercial outfit to dock its own cargo capsule at the ISS. "It looks like we got us a Dragon by the tail," said U.S. astronaut Don Pettit, who was operating the Canadarm2, the Canadian-built robotic arm, as it reached out and hooked on to Dragon at 9:56 am (13:56 GMT). Experts immediately hailed the historic flight as a new era for private spaceflight and NASA officials agreed. While attached to the ISS, Dragon delivered 305 kilograms of food, clothing and other supplies, as well as 122 kilograms of cargo bags, 20 kilograms of science experiments, and 10 kilograms of computer equipment.

After spending a few days at the orbiting outpost, the Dragon made its way home after detaching from the robotic arm. Five hours later, the Dragon used its thrusters to begin its de-orbit burn 400 kilometers above the Indian Ocean, finally splashing down into the Pacific Ocean several hundred kilometers off the coast of Baja California, at 8:42 a.m. PDT. The capsule was recovered by boats and brought to the port of Los Angeles. From launch to the splashdown, the Dragon mission had lasted 9 days, 7 hours and 58 minutes. Dragon had joined Russia's Progress, the ESA's Automated Transfer Vehicle (ATV) and the Japan Aerospace Exploration Agency's H-II Transfer Vehicle (HTV) as regular station suppliers. However, Dragon was the only provider in the international lineup with the capability of returning to Earth a significant cargo of research samples and equipment in need of refurbishment, a critical part of future science activities planned for the orbiting science laboratory.

The successful third Falcon 9/Dragon flight was not the only win for SpaceX that week. Just days after Dragon docked with the ISS, it was announced that SpaceX and satellite service provider Intelsat had reached a commercial agreement for the launch of a Falcon Heavy rocket, which is discussed later in this chapter. But before moving on to the Falcon Heavy, we need to follow the timeline of the Falcon 9. The first Falcon 9 – the Falcon 9 v1.0 – was eventually launched five times between June 2010 and March 2013. It was during the lifetime of this launch vehicle that SpaceX began its development of a reusable Falcon 9, of which more shortly. Taking over from the Falcon 9 v1.0 was the Falcon 9 v1.1, which flew 15 times between September 2013 and January 2016. This vehicle was 60 percent heavier than its predecessor and also 60 percent more powerful. Powered by nine Merlin 1D engines arranged in the SpaceX *Octaweb* configuration (a cluster of eight engines surrounding a ninth, central one), the Falcon v1.1 delivered 5,885 kN and could carry 9,000 kilograms to LEO.

Following on from Falcon 9 v1.1 was the Falcon 9 Full Thrust. Featuring cryogenic cooling of its propellant to increase propellant density, the Full Thrust iteration also features a reusable first stage, which achieved its first successful landing in December 2015. This booster was developed to facilitate quick reusability. The testbeds for this technology were the SpaceX Grasshopper and Falcon 9 Reusable Development (F9R Dev) vehicles. The Grasshopper project was an internally funded program announced in 2011. The vehicle was 32 meters tall and made eight low-altitude hover-landing test flights between 2012 and 2013 (see Table 5.5). This development experience formed the basis for a larger and more capable prototype, the F9R Dev, which tested at higher altitudes and higher speeds. This vehicle made its first flight test in April 2014 but was subsequently destroyed in an accident four months later. The Grasshopper and F9R Dev tests laid the foundation for the development of the reusable Falcon 9, which ultimately led to that first booster landing on December 21, 2015. Announced in October 2012, the F9R Dev, which was originally dubbed Grasshopper v1.1, was constructed with four retractable landing legs and was designed to extend the performance envelope of its predecessor. It did so successfully during its short, five-flight test program (see Table 5.6). Prior to the loss of F9R Dev, SpaceX had been planning a third test vehicle – the F9R Dev2 – but since the F9R Dev version had mostly succeeded in achieving the test flight objectives, SpaceX abandoned that idea.

Table 5.5: Grasshopper Test Flight Summary

Test #	Date	Altitude	Duration	Remarks
1	2012 Sep 21	1.8 m	3 sec	A brief hop with a near-empty tank.
2	2012 Nov 01	5.4 m	8 sec	
3	2012 Dec 17	40 m	29 sec	First flight to include cowboy mannequin.
4	2013 Mar 07	80 m	34 sec	Touchdown thrust-to-weight ratio > 1.
5	2013 Apr 17	250 m	58 sec	Demonstrated ability to maintain stability in wind.
6	2013 Jun 14	325 m	68 sec	New navigation sensor suite tested.
7	2013 Aug 13	250 m	60 sec	Performed 100 m lateral maneuver.
8	2013 Oct 07	744 m	79 sec	Final flight.

Table 5.6 F9R Dev Test Flight Summary

Test #	Date	Test vehicle	Highest altitude	Remarks
1	2014 Apr 17	F9R Dev1	250 m	Hovered, moved sideways, landed successfully.
2	2014 May 01	F9R Dev1	1,000 m	Hovered, moved sideways, landed.
3	2014 Jun 17	F9R Dev1	1,000 m	First test flight with steerable grid fins.
4	2014 Aug 01	F9R Dev1		No information provided by SpaceX.
5	2014 Aug 22	F9R Dev1		Vehicle exploded following a flight anomaly caused by blocked sensor.

With the success of their Grasshopper and F9R Dev programs, SpaceX moved to their Falcon 9 vehicle for Vertical Takeoff, Vertical Landing (VTVL) tests. This next phase of VTVL testing culminated in the successful landing of the Falcon 9 Flight 20 on December 21, 2015. SpaceX achieved the first re-flight of the first stage 18 months later, and since then has made incremental upgrades to the reusability of its Falcon 9 Full Thrust vehicle. By May 2016, SpaceX had completed three first-stage landings on their drone ship (Figures 5.4 and 5.5).

Figure 5.4. Falcon 9 first stage (from the CRS-6 flight in 2015) about to land on an Autonomous Spaceport Drone Ship (ASDS – a modified barge). In this image, the landing legs are deploying. Credit SpaceX.

Further modifications to the Falcon 9 Full Thrust came in 2017 when SpaceX made minor engine thrust upgrades, the result of which was designated as the Block 4 version. This version first flew on the CRS-12 mission on August 14, 2017. The design was improved again the following year to become the Block 5 version (described as the final version of the vehicle by Musk), which featured second stage upgrades that enabled the vehicle to operate longer on orbit, a capability achieved by multiple reignitions of the second stage engine. The Block 5 version first flew on May 11, 2018.

Figure 5.5. A closer look at the CRS-6 first stage about to land on the ASDS. Credit SpaceX.

In terms of reliability, at the time of writing, the Falcon 9 has operated at 98.4 percent success rate (124 out of 126 launches), which makes it the most successful launch vehicle in operation (the Soyuz has a 95.1 percent success rate after 1,880 launches). Its launch sequence, summarized in Table 5.7, includes that aforementioned hold-down feature, similar to the one employed by the Shuttle, in which the vehicle is held down after the first stage starts until all systems are confirmed to be operating nominally. Another layer of safety is provided by the Falcon 9's engine-out capability. This feature was first utilized during the CRS-1 mission in October 2012 when, during the climb to orbit, engine Number 1 lost pressure at T+79 seconds and shut down. To overcome the loss of acceleration, the first stage simply burned for an additional 28 seconds and the second stage had to burn for an extra 15 seconds. A similar situation occurred during the March 2020 Starlink mission, when one of the first stage engines failed three seconds before main engine cut-off.

Table 5.7: Falcon 9 Launch Profile

Time	Event
T-0:00:03	Merlin Engine Ignition
T-0:00:00	Liftoff
T~0:00:16	Pitch Kick & Roll Maneuver
T~0:01:08	Mach 1
T+0:01:22	Maximum Dynamic Pressure, Throttle Down
T+0:02:38	MECO – First Stage Cut Off
T+0:02:41	Stage Separation
T~0:02:44	1st Stage Maneuver out of 2nd Stage Plume
	1st Stage Re-Orientation to Engines first
T+0:02:49	Mvac Ignition
T+0:03:49	Payload Fairing Jettison
T+0:06:19	1st Stage Re-Entry Burn (~25 sec)
T+0:08	1st Stage Landing Burn
T~0:08:32	1st Stage Landing
T+0:08	2nd Stage Terminal Guidance Mode
T+0:08	2nd Stage Flight Termination System safe
T+0:08:34	2nd Stage Engine Shutdown
T+0:26:29	Second Stage Re-Start
T+0:27:22	Second Stage Cut Off

In addition to being reliable, the Falcon 9 is also reusable (Table 5.8). In the early days, SpaceX attempted recovery of the first stages using parachutes, but the stages did not survive the aerodynamic stress of re-entry, so the company did away with the parachutes and switched to retro propulsion, a design that was completed in February 2012 (Table 5.9). Thanks to experience gained when testing the technology using the Grasshopper – and a few failed landings – SpaceX eventually pronounced the landing attempts to be a routine procedure. Following that announcement, the next goal was the operational reuse of a previously flown booster, a feat accomplished in March 2017. Four years later, SpaceX has now re-flown one booster ten times. Still, reusing the first stage is just one part of the reusability puzzle. There is still the business of recovering and reusing the fairings (which cost $6 million) and the second stage. Space X recovered a fairing from the SES-10 mission in March 2017, but they have not had much success recovering the second stages because the development and design work to achieve this goal has been transferred to the Starship. Thanks to this reusability, SpaceX is able to maintain its pricing on the low side, with a standard Falcon 9 base going for $52 million in 2021. These prices are so low that the company's competitors (Ariane, United Launch Alliance, International Launch Services, Roscosmos) have accused Musk's company of price-dumping. Of course, the main reason the Falcon 9 flights are so cheap is that the vehicle is 80 percent reusable, whereas the other vehicles in the game remain expendable.

Table 5.8: Notable Falcon 9 Flights

Flight 1	**Dragon Spacecraft Qualification Unit**	1st flight of Falcon 9 and 1st test of Dragon.
Flight 3	**Dragon C2+**	1st cargo delivery to ISS.
Flight 6	**Cassiope**	1st v1.1 rocket, 1st launch from Vandenberg AFB, 1st attempt at propulsive return of the first stage.
Flight 7	**SES-8**	1st launch to Geosynchronous Transfer Orbit (GTO), 1st commercial payload (communications satellite).
Flight 9	**SpaceX CRS-3**	Added landing legs, 1st fully controlled descent and vertical ocean touchdown.
Flight 15	**Deep Space Climate Observatory (DSCOVR)**	1st mission passing escape velocity to the L1 point.
Flight 19	**SpaceX CRS-7**	Total loss of mission due to structural failure and helium overpressure in the second stage
Flight 20	**Orbcomm OG-2**	1st vertical landing of an orbital-class rocket.
Flight 23	**SpaceX CRS-8**	1st landing vertically achieved on an autonomous spaceport drone ship at sea.
(Flight 29)	**Amos-6**	Total vehicle and payload loss prior to static fire test.
Flight 30	**SpaceX CRS-10**	1st launch from LC-39A at the Kennedy Space Center.
Flight 32	**SES-10**	1st re-flight of a previously flown orbital class booster (B1021, previously used for SpaceX CRS-8), 1st recovery of a fairing.
Flight 41	**Boeing X-37B OTV-5**	1st launch of a spaceplane.
Flight 54	**Bangabandhu Satellite-1**	1st flight of the Block 5 version.
Flight 58	**Telstar 19V**	Heaviest communications satellite ever delivered to GEO.
Flight 69	**Crew Dragon Demo-1**	1st launch of the Crew Dragon.
Flight 72	**RADARSAT Constellation**	The most valuable commercial payload put into orbit.
Flight 85	**Crew Dragon Demo-2**	1st crewed launch of the Crew Dragon.
Flight 98	**Crew-1**	1st crewed operational launch of the Crew Dragon, holding the record for the longest spaceflight by a U.S. crew vehicle.
Flight 101	**SpaceX CRS-21**	1st launch of the Cargo Dragon 2, an uncrewed variation of the Crew Dragon.
Flight 106	**Transporter-1**	1st dedicated small sat rideshare launch, setting the record of the most satellites launched on a single launch at 143. Surpassed the previous record of 108 satellites held by the November 17, 2018 launch of an Antares.

"After that it's just going to be out there in space for maybe millions or billions of years, maybe discovered by some future alien race thinking what the heck, what were these guys doing? Did they worship this car?"

Elon Musk, speaking at a SpaceX press conference shortly after launching his Tesla Roadster on board a Falcon Heavy.

Table 5.9: Falcon 9 Specifications

Version	v1.0 (retired)	v1.1 (retired)	v1.2 or Full Thrust	
			Block 3 and Block 4 (retired)	Block 5 (active)
Stage 1 engines	9 × Merlin 1C	9 × Merlin 1D	9 × Merlin 1D (upgraded)	9 × Merlin 1D (upgraded)
Stage 1 mass (t)			dry mass 22.2 t	
Stage 2 engines	1 × Merlin 1C Vacuum	1 × Merlin 1D Vacuum	1 × Merlin 1D Vacuum (upgraded)	1 × Merlin 1D Vacuum (upgraded)
Stage 2 mass (t)			dry mass 4 t	
Height (m)	53	68.4	70	70
Diameter (m)	3.66	3.66	3.66	3.66
Initial thrust	3.807 MN (388.2 tf)	5.9 MN (600 tf)	6.804 MN (693.8 tf)	7.6 MN (770 tf)
Takeoff mass (tonnes)	318	506	549	549
Fairing diameter (m)	N/A	5.2	5.2	5.2
Payload to LEO (kg) (from Cape Canaveral)	8,500–9,000	13,150	22,800 (expendable)	≥ 22,800 (expendable) ≥ 16,800 (reusable)
Payload to GTO (kg)	3400	4850	8300 (expendable) About 530 (reusable)	≥ 8300 (expended) ≥ 5,800 (reusable)
Success ratio	5 / 5	14 / 15	36 / 36 (1 precluded)	67 / 67

The Falcon Heavy comprises a strengthened center core Falcon 9 first stage with two Falcon-9 strap-on boosters either side. Not surprisingly, at the time of writing, it is the most powerful operational launch vehicle. The vehicle's maiden launch on February 6, 2018, was quite an event thanks to the unusual payload (Figure 5.6), which caused quite a stir among the Twitterati and the Musk faithful – or Musketeers, as they are occasionally referred to. The Falcon Heavy (Figure 5.7) launched for the second time in April 2019 and again in June of the same year. Thanks to its success and payload capability, it is unsurprising that the vehicle has a healthy manifest stretching into the mid-2020s (see Table 5.10). Having said that, what may be surprising to some is that the vehicle is living on borrowed time. That is because the Falcon 9 and Falcon Heavy will eventually be superseded by the do-it-all Starship. It is one of the reasons SpaceX is not pursuing the human-rating certification for the Falcon Heavy.

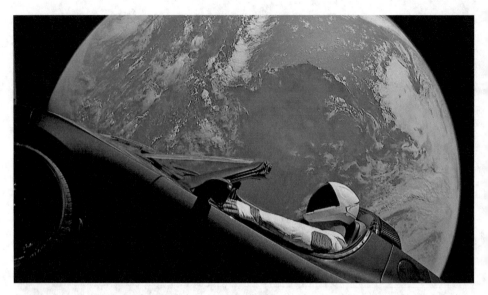

Figure 5.6. Elon Musk's Spaceman mannequin seated in the driver's seat of the Mars Messiah's Tesla Roadster. Credit SpaceX.

Figure 5.7. Falcon Heavy maiden launch, February 6, 2018. The vehicle can lift 63,800 kilograms into low Earth orbit (LEO) and 26,700 kilograms to geostationary transfer orbit (GTO). The first stage comprises three Falcon 9 cores, each powered by nine Merlin engines. The upper stage is powered by one Merlin engine. Following separation of the side boosters, the center core throttles to maximum thrust. Each core features four extensible landing legs and grid fins, which are deployed after stage separation. Following stage separation, the center engine in each core burns for a few seconds to control the trajectory to facilitate landing of the cores on the drone ship. Credit SpaceX.

The Falcon Heavy concept stretches back to a SpaceX news update in September 2005. Thirteen years later, the Falcon Heavy made its first flight, at a development cost of just $500 million (peanuts in the launch industry – by comparison, NASA has spent more than $20 billion developing the Space Launch System which is still yet to launch). Although actual testing did not begin in 2013, there is a reason for this; SpaceX first needed to have a solid understanding of the capabilities and performance of the Falcon 9. That level of understanding was achieved once the company had flown a few Falcon 9 flights, and it led to what was a rapid development and testing program for the Falcon Heavy. A significant part of that testing took place at the SpaceX Rocket Development and Test Facility in McGregor, Texas. It was here that SpaceX tested the triple cores and their 27 rocket engines.

Musk downplayed expectations in advance of the Falcon Heavy's maiden flight, stating there was a good chance the vehicle would not make orbit. Fortunately, the flight went very well, and on that February 6, the first Falcon Heavy lifted off, carrying Musk's Tesla Roadster into an orbit around the Sun. Thanks to three strategically positioned cameras, the media was treated to some rather unique images of a suited mannequin at the wheel of a sportscar in space. Shortly after the Roadster flight, SpaceX signed contracts worth more than $500 million, which meant that the development cost of the vehicle had been covered. There was also discussion about using the Falcon Heavy as an alternative to the obscenely expensive Space Launch System (SLS). While there was a lot of support for this among the Musketeers, the Falcon Heavy could not replace the SLS because although it is powerful enough to launch

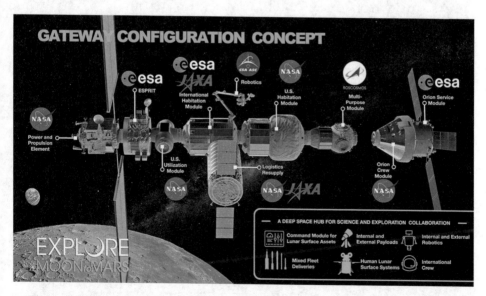

Figure 5.8. The Gateway is a key element of NASA's Artemis Program. Designed to be an outpost orbiting the Moon, the Gateway comprises several modules, two of which – the PPE and HALO – will be launched on Falcon Heavy in 2024. The PPE is a 60-kilowatt module that will provide power, attitude control and communications capabilities for the Gateway, while the HALO will serve primarily as a crew cabin. Credit NASA.

the Orion, it cannot carry the Orion *and* the European Service Module. Having said that, the Falcon Heavy will be supporting Artemis missions (Figure 5.8), as the vehicle has been selected to launch the Power and Propulsion Element (PPE) and the Habitation and Logistics Outpost (HALO) elements of the Lunar Gateway in 2024.

Table 5.10: Falcon Heavy Specifications

Characteristic	First stage core unit (1 × center, 2 × booster)	Second stage	Payload fairing
Height	42.6 m	12.6 m	13.2 m
Diameter	3.66 m	3.66 m	5.2 m
Dry Mass	22.2 t	4 t	1.7 t
Fueled mass	433.1 t	111.5 t	N/A
Structure type	LOX tank: monocoque Fuel tank: skin and stringer	LOX tank: monocoque Fuel tank: skin & stringer	Monocoque halves
Structure material	Aluminum–lithium skin; aluminum domes	Aluminum–lithium skin; aluminum domes	Carbon fiber
Engines	9 × Merlin 1D	1 × Merlin 1D Vacuum	
Propellant	Subcooled liquid oxygen, kerosene (RP-1)	Liquid oxygen, kerosene (RP-1)	
LOX tank capacity	287.4 t	75.2 t	
Kerosene tank capacity	123.5 t	32.3 t	
Thrust, stage total	22.82 MN sea level	934 kN vacuum	
Propellant feed system	Turbopump	Turbopump	N/A
Throttle capability	419–816 kN sea level	360–930 kN vacuum	
Restart capability	3 engines for boost-back, reentry, and landing	Dual redundant TEA-TEB pyrophoric igniters	
Ascent attitude control: pitch, yaw	Gimbaled engines	Gimbaled engine and nitrogen gas thrusters	
Ascent attitude control: roll	Gimbaled engines	Nitrogen gas thrusters	
Coast/descent attitude control	Nitrogen gas thrusters and grid fins	Nitrogen gas thrusters	Nitrogen gas thrusters
Stage separation system	Pneumatic	N/A	Pneumatic

Prices for a Falcon Heavy launch have varied as vehicle development progressed (see Table 5.11). Initially, in 2011, a Falcon Heavy launch was advertised as $80 to $125 million, a number that increased to $150 million for 63.8 tonnes to LEO ($2,350 per kilogram to LEO). In comparison, a Delta IV Heavy with a payload of 28.4 tonnes to LEO will cost you $12,340 per kilogram.

Table 5.11: Falcon Heavy launches

Flight No.	Launch date	Payload and mass	Customer	Price	Outcome
1	2018 Feb 6	Elon Musk's Tesla Roadster ~1,250 kg	SpaceX	Internal	Success
2	2019 Apr 11	Arabsat-6A 6,465 kg	Arabsat	Undisclosed	Success
3	2019 Jun 25	USAF STP-2 3,700 kg	DoD	US$160.9 million	Success
–	Early 2022	USSF-44	U.S. Space Force	US$130 million	Scheduled
–	Early 2022	USSF-52	U.S. Space Force	US$99 million	Scheduled
–	Q1 2022	ViaSat-3 Americas	Viasat		Planned
–	2022	Inmarsat-6B	Inmarsat		TBA
–	August 2022	*Psyche*	NASA (Discovery)	US$117 million	Scheduled
–	Q3 2022	USSF-67	USSF	US$317 million	Planned
–	November 2023	Griffin Mission 1	Astrobotic/NASA (Artemis)	Undisclosed (list price US$90 million)	Planned
–	October 2024	*Europa Clipper*	NASA (Planetary Missions)	US$178 million	Planned
–	November 2024	Power and Propulsion Element (PPE) Habitation and Logistics Outpost (HALO)	NASA (Artemis)	US$331.8 million	Planned
–	2024	At least two Dragon XL flights	NASA (Gateway Logistics Services)		Planned
–	TBA	TBA	Intelsat		TBA

6

The Dragon has Landed

Picking up where NASA left off

"Looks like we got us a dragon by the tail."

Astronaut Don Pettit as he extended the International Space Station's robotic arm and captured the Dragon capsule.

Figure 6.0. The Dragon cargo vehicle is grappled by the International Space Station's robotic arm. Working from the robotics workstation inside the Cupola, Aki Hoshide and Sunita Williams captured Dragon on October 10, 2012. Credit NASA.

© Springer Nature Switzerland AG 2022
E. Seedhouse, *SpaceX*, Springer Praxis Books,
https://doi.org/10.1007/978-3-030-99181-4_6

On May 31, 2012, a cargo-laden SpaceX Dragon parachuted back to Earth after a nearly flawless demonstration mission to the International Space Station (ISS). After splashing down in the Pacific Ocean about 800 kilometers west of Baja California, the unmanned capsule was retrieved by recovery ships before being hauled to the Port of Los Angeles and transported overland to the SpaceX processing facility in McGregor, Texas, where it underwent final inspection. The landing capped a successful nine-day mission that saw Dragon become the first commercial vehicle to visit the ISS, thereby setting the stage for regular cargo missions to the orbiting outpost. The first of these took place in October 2012 (Figure 6.0). In addition to triggering a series of 12 cargo missions NASA had ordered from SpaceX, the Dragon's landmark mission proved that uncrewed cargo vessels sent to the ISS could be recovered and reused. The flight also boosted the company's efforts to make its spacecraft suitable for crewed missions. For SpaceX (a contractor relatively independent of NASA) to have developed, built, tested, *and* flown its spacecraft so successfully, the flight ranked near the top of the list of firsts in the U.S. space program. Dragon's demonstration mission also marked NASA's return to flight following the retirement of the Shuttle in 2011 and dispelled some of the doubts among members of Congress, some of whom had voted to cut President Obama's $830 million budget request for commercial spacecraft development. Under NASA's Commercial Resupply Services (CRS) program, SpaceX was contracted to deliver cargo to the ISS beginning in October 2012. The company was also awarded a Commercial Crew Development (CCDev) contract in April 2011 to carry astronauts, or a combination of personnel and cargo, to and from Low Earth Orbit (LEO). The development of the crewed version of the Dragon will be discussed in the following chapter, but for this chapter we will focus our attention on the cargo variant.

Figure 6.1. Dragon leaving SpaceX headquarters in Hawthorne, California, February 2015. Credit SpaceX. Public domain.

Dragon's development started in late 2004 when SpaceX initiated the design using its own funding. The capsule formed the centerpiece of the proposal SpaceX submitted under NASA's Commercial Orbital Transportation Services (COTS) program. The gumdrop-shaped Dragon (Figure 6.1) comprises a blunt-cone ballistic capsule not dissimilar to the design of the Soyuz or Apollo capsules, a nose-cone cap that jettisons after launch, and a trunk equipped with two solar arrays. To protect its cargo during re-entry, the capsule utilizes a proprietary variant of NASA's Phenolic Impregnated Carbon Ablator (PICA) material. The spacecraft also sports standard options such as a docking hatch, maneuvering thrusters – 18 of them – and the trunk, which, unlike the rest of the re-usable spacecraft, separates from the capsule before re-entry. Dragon is launched atop the Falcon 9 booster and its mode of landing, which has been very successful, is to splash down in the Pacific Ocean before being returned to shore by ship.

Before we describe the Dragon's development and design, there may be some reading this who are wondering how the spacecraft got its name. The rule when it comes to naming SpaceX rockets or rocket parts, or anything else in the Mars Messiah's universe, is that the names must be *cool*. In the world of SpaceX, these names also tend to have some cinematic significance. As previously mentioned, for example, the Falcon and its siblings were named after the *Millennium Falcon* flown by Han Solo in the sci-fi blockbuster *Star Wars*. The Merlin engines (in addition to the falconry derivation) may well be named after the wizard Merlin. The business of using cool monikers does not stop there; a navigation sensor dubbed DragonEye was tested in 2009 on *Endeavour*'s STS-127 mission when the Shuttle was approaching the ISS. Then there are the thrusters that Dragon uses to maneuver in orbit. Dragon features 'Draco' thrusters and, as any Harry Potter fan knows, 'Draco' is one of the bad guys in the series of films. The 'Kestrel' moniker, a name given to the engines that powered the upper stage of the Falcon 1 rocket, is no coincidence either, since kestrels happen to be the name of a bird in the falcon genus. As for the Dragon itself, this spacecraft is named after the fictional creature "*Puff the Magic Dragon*" in the song by Peter, Paul and Mary. Why? It was in response to widespread critical opinion, when Musk started his company, that his goals were out of reach. The naming trend SpaceX uses differs markedly from that of NASA, whose spacecraft currently in development is known by its acronym MPCV, for Multi-Purpose Crew Vehicle, although the spacecraft tends to go by the name Orion. The agency's creativity in the naming department did not get much better when it came to thinking up an appellation for a certain $150 billion orbiting outpost you may have heard of, known simply as the International Space Station.

In 2005, the year after SpaceX began developing Dragon, NASA initiated its COTS development program, soliciting proposals for commercial resupply spacecraft to replace the soon-to-be retired Space Shuttle. SpaceX submitted Dragon as part of its proposal in March 2006 and six months later, on August 18, 2006, NASA announced SpaceX had been chosen, along with Kistler Aerospace, to develop ISS cargo launch services. The initial plan called for three Dragon demonstration flights to be flown between 2008 and 2010, for which SpaceX would receive up to $278 million if they met NASA's milestones[1]. Two years later, on December 23, 2008, NASA awarded a $1.6 billion CRS contract to SpaceX, with options to increase the contract value to $3.1 billion. The contract called for 12 cargo flights carrying a minimum of 20,000 kg to the ISS. Dragon's development progressed quickly. On February 23, 2009, SpaceX announced that Dragon's PICA-X heat shield material had passed heat stress tests in preparation for the capsule's maiden launch. Then, in July 2009, the capsule's primary proximity-operations sensor, DragonEye, was tested during the aforementioned STS-127 mission. A later Shuttle mission, STS-129, delivered the COTS UHF Communication Unit (CUCU) and Crew Command Panel (CCP); the CUCU allows the ISS to communicate with Dragon, while the CCP allows ISS crew members to issue basic commands to the capsule. These communication units and associated hardware are important for conducting proximity operations, which we will discuss later in this chapter.

Dragon's first flight took place in June 2010, when a stripped-down capsule dubbed the Dragon Spacecraft Qualification Unit (SQU – initially used as a ground testbed) was launched on a Falcon 9. Dragon's first mission was simply to relay aerodynamic data captured during the ascent, because the capsule was not designed to survive re-entry since SpaceX did not have a re-entry license, a document issued by the Federal Aviation Administration (FAA). Following the successful flight of the first Dragon, SpaceX then had to wait for the FAA to issue that re-entry license, which the agency did on November 22, 2010, making it the first such license awarded to a commercial vehicle. On December 8, 2010, less than three weeks after the re-entry license had been granted, the first Dragon spacecraft launched on COTS Demo Flight 1, which also marked the second flight of the Falcon 9. Unlike the SQU, this Dragon *was* designed to survive re-entry and was successfully recovered (Table 6.1).

Following the recovery of the second Dragon, SpaceX continued its preparations for COTS 2, including further on-orbit testing of the DragonEye sensor which flew again on STS-133 in February 2011. Then, on April 18, 2011, SpaceX was awarded $75 million of funding during the first phase of NASA's Commercial Crew Development (CCDev) milestone-based program to help develop the agency's crew system, as discussed in the following chapter. Under the CCDev program, the milestones for SpaceX included: the advancement of the Falcon 9/Dragon crew transportation design; the development and testing of the Launch Abort System (LAS) propulsion; completion of two crew accommodations demonstrations; and full-duration test firings of the launch abort engines and demonstrations of their throttle capability. At the end of the year, thanks to the success of the Falcon 9 and its weyr of Dragons (a fictional term used to describe a group of dragons), NASA approved the decision by SpaceX to combine the COTS 2 and 3 mission objectives into one Falcon 9/Dragon flight, designated as COTS 2+ or COTS 2/3. The mission was required to meet a number of objectives (Figure 6.2).

---

[1] Kistler's contract was terminated in 2007 after it failed to meet its obligations and NASA re-awarded Kistler's contract to Orbital Sciences.

Figure 6.2. Demonstration mission objectives. Credit NASA.

Table 6.1: Dragon Specifications

Characteristic			Characteristic	
Dry Mass	4201 kilograms		Launched	23
Payload	6000 kilograms upmass 3000 kilograms downmass		Maiden flight	Dec 8, 2010
Volume	10 m^3 pressurized 14^3 unpressurized		Derivatives	DragonLab Dragon 2 Dragon XL Red Dragon
Length	6.1 meters		Applications	ISS Logistics
Diameter	3.7 meters		Propellant	NTO / MMH

Flight Day 1 of the C2/3 mission began with vehicle power-up and countdown operations, including the final rounds of testing of both the Falcon 9 and Dragon. This phase included fueling the launch vehicle and configuring Dragon for on-orbit operations. The Terminal Countdown Sequence was initiated ten minutes before lift-off, during which the final configurations of each vehicle were set by computer commands. The Falcon 9's nine Merlin 1C engines ignited three seconds before launch, and the vehicle lifted off from Space Launch Complex (SLC) 40 at Cape Canaveral Air Force Station (CCAFS), Florida, at T-0. After completing a short vertical ascent, the Falcon 9 made its now familiar roll and pitch maneuver to align itself with the correct flight trajectory (Table 6.2). The vehicle reached maximum dynamic pressure (Max Q) 84 seconds after launch, and the first stage shut down and separated from the vehicle three minutes into the mission. Seven seconds after stage separation, the second stage's Merlin ignited, beginning a six-minute burn. During the second stage burn, the Dragon's nose cone separated to increase ascent performance, exposing the Spacecraft Docking System (SDS). The vehicle shut down its engine nine minutes and 14 seconds after launch, and the Dragon was released 35 seconds later. Shortly thereafter, Dragon deployed its solar arrays, while SpaceX Dragon Control, based in Hawthorne, California, began its list of vehicle status checks to confirm the capsule was in good condition and ready for on-orbit operations.

Table 6.2: Post-launch events

MET[1]	Event
T-0:00:03	Merlin Engine Ignition
T+0:01:24	Maximum Dynamic Pressure – Max Q
T+0:03:00	First Stage Cut-off
T+0:03:05	Stage Separation
T+0:03:12	Second Stage Ignition
T+0:03:52	Dragon Nose Cone Jettison
T+0:09:14	Second Stage Cut-off
T+0:09:49	Spacecraft Separation

[1] Mission Elapsed Time

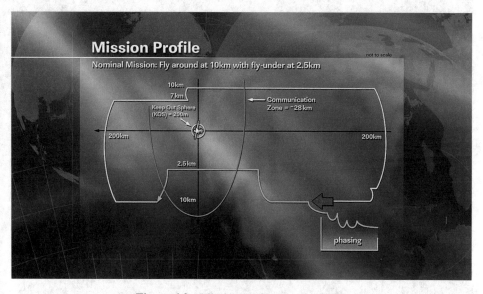

Figure 6.3. Mission profile. Credit NASA

Having successfully made it to orbit, the next task for Dragon was to approach the ISS and dock with it. This is a tricky part of such missions, because rendezvousing and docking a spacecraft is fiendishly challenging. To give the reader an idea of the complexities involved, there follows a series of graphics that are supported by events listed in corresponding tables. Some of these events will be addressed here. Dragon's first day in space was dedicated to Far Field Phasing

Maneuvers (Figure 6.3) and various test objectives. As the mission progressed, Dragon was required by NASA to complete a series of tests and meet several objectives before being permitted to rendezvous with the ISS (Figures 6.4 and 6.5). One of the first of these tests was to demonstrate its Global Positioning System (GPS) navigation capability, a task that was completed about an hour into the flight. Shortly after succeeding with that test, Dragon deployed its Guidance, Navigation and Control (GNC) bay door to position the necessary rendezvous instruments. The vehicle's DragonEye rendezvous and navigation instrument suite included a Light Detection and Ranging (LIDAR) imager for this purpose. Shortly after being deployed, each of these instruments was activated and underwent their requisite checkouts (Table 6.3).

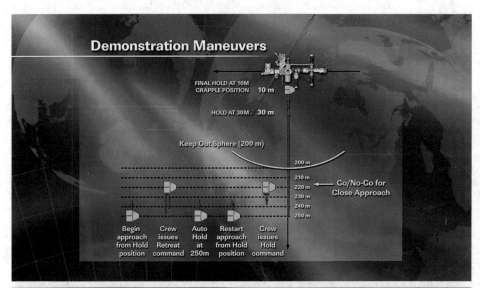

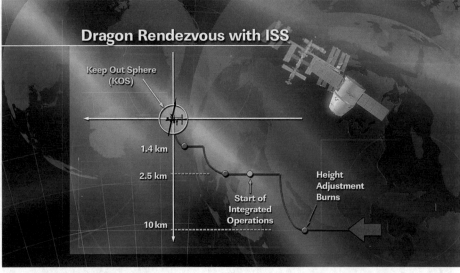

Figures 6.4 and 6.5. Mission objectives. Credit NASA

Table 6.3: Initial Orbit Operations Timeline

MET[1]	Event
00/00:11:53	Solar Array Deployment
00/00:54:49	Absolute GPS Demonstration
00/02:26:48	GNC Bay Door Deployment
00/02:40:49	Relative Navigation Sensors Checkout (LIDAR, Thermal)
00/04:11:20	Orbit Adjust Burn
00/08:46:52	Full Abort Test (Continuous Burn)
00/09:31:25	PCE1 Burn
00/09:57:58	Pulsed Abort Test
00/10:37:58	Free Drift Demonstration

[1.] Mission Elapsed Time

On Flight Day 2, Dragon made several engine burns to adjust its orbit in preparation for its rendezvous with the ISS the following day. The burns, referred to as
Far Field Phasing or Height Adjust Burns in NASA parlance, were conducted to
increase Dragon's orbital altitude. This required the vehicle to target a point 10
kilometers beneath the ISS. The Dragon's rendezvous profile from orbit insertion
to docking at the ISS can be divided into three phases: far-field, mid-field and
proximity operations. The far-field stage is characterized as the most quiescent
phase. En-route to the ISS, the vehicle uses Inertial Measurement Units (IMUs) to
calculate its position and uses this opportunity to take extensive ground-based
radar updates. The targeting solutions for burn maneuvers are also computed on
the ground and uplinked. The mid-field rendezvous phase starts as the Dragon
utilizes relative sensors for onboard navigation, a process coordinated through
timelines anchored to various mission events. The final phase of rendezvous is
proximity operations, which involves the frequent use of maneuvering thruster
firings to control the relative trajectory of the vehicle up to docking.

Once Dragon had arrived at a point 10 kilometers below and behind the ISS
(red arrow in Figure 6.4), teams at each Mission Control Center conducted a poll
before permitting Dragon to conduct its Fly-Under Maneuver to set it up in preparation for performing the R-Bar Maneuver (RBM)[2]. Once approved, Dragon performed two burns called HA2 and CE2 before entering the space station's
28-kilometer communications zone, within which it could communicate directly

---

[2]The RBM was a maneuver performed by the Shuttle as it rendezvoused with the ISS. The
Shuttle performed a backflip that exposed its heatshield to the ISS crew who took photographs
of it. Based on analysis of the images, mission control could decide if the orbiter was safe for
re-entry (this became a standard procedure following return to flight after the *Columbia* accident, which had been caused by a damaged heat shield). The name of the maneuver was based
on the R-bar and V-bar lines used in the approach to the ISS. R-bar, or Earth Radius Vector, is
an imaginary line connecting the ISS to the center of the Earth. The RBM was developed by
NASA engineers Steve Walker, Mark Schrock and Jessica LoPresti following the *Columbia*
disaster in February 2003.

with the ISS. For relative GPS communication with the ISS, Dragon used a proximity communications link supplemented by the COTS Ultra High Frequency (UHF) Communication Unit (CUCU) aboard the station. As it made its close approach, Dragon demonstrated its Relative GPS System (RGPS), which determined the spacecraft's position relative to the ISS. During the test, the RGPS was assessed by comparing the data it obtained with simultaneously acquired absolute GPS data. To test the CUCU, ISS astronauts Andre Kuipers and Don Pettit, who were responsible for Dragon operations (Figure 6.6), sent a strobe/test command to Dragon. When Dragon received and executed the command, it activated a light for visual confirmation of successful CUCU communications.

Once Dragon had passed 2.5 kilometers below the ISS and all operations were complete, the vehicle made a trajectory adjustment to retreat to 10 kilometers to begin the ISS fly-around. For the fly-around maneuver, the vehicle made several engine burns to cross the Velocity Vector (V-bar) of ISS, flying to a location seven kilometers above the V-Bar before making another maneuver to reduce its velocity. Dragon then passed over the ISS, making one more height adjustment burn to increase the distance between itself and ISS to 10 kilometers. Dragon once again fired its engines to cross the V-Bar one more time, targeting a point behind and below the ISS. The sequence of events took a full day and finally set the stage for re-rendezvous. The data acquired during the Fly-Under (Table 6.4) was reviewed by SpaceX and NASA teams, after which the ISS Mission Management Team made a Go/No Go decision for rendezvous.

Figure 6.6. Capturing a Dragon. Don Pettit (L) and Andre Kuipers practice robotic workstation operations in preparation for the Dragon's arrival. Credit NASA.

Table 6.4: Fly-Under Timeline

MET	Event
01/23:18	Height Adjustment Burn #2
02/00:04	Co-Elliptic Burn #2
02/00:15	Relative GPS Demonstration
02/00:54	Entered ISS Communication Zone
02/02:44	Crossed R-Bar at 2.5 kilometers
02/03:23	Departure Burn #1
02/04:10	Departure Burn #2
02/06:47	Forward Height Adjust Burn #1
02/07:33	Forward Height Adjust Burn #2
02/12:14	Forward Co-Elliptic Burn #2
03/16:28	Rear Height Adjust Burn #1

During maneuvering, Dragon was restricted to the following:
Closing (axial rate): 0.05 to 0.10 meters per second
Lateral (radial) rate: 0.04 meters per second
Pitch/yaw rate: 0.15° per second (vector sum of pitch/yaw rate)
Roll rate: 0.40° per second
Lateral (radial) misalignment: 0.11 meters
Pitch/yaw misalignment: 5 degrees (vector sum of pitch/yaw rate)
(Source: International Space Station partnership)

Several Dragon systems were checked at the 250-meter hold, including rendez-vous navigation systems and LIDAR, to demonstrate DragonEye's capabilities. Once Mission Control verified that Dragon's position and velocity was accurate, the approach re-started and Dragon made short engine pulses to re-initiate the rendezvous. Once the spacecraft reached 220 meters on the R-Bar, the ISS crew sent a retreat command to demonstrate one of Dragon's rendezvous abort capabilities. Once this had been done, Dragon fired its engines again to return to the 250-meter hold point, a maneuver it would be required to perform at any stage during the approach if a retreat command was sent. While Dragon completed the retreat operation, Mission Control assessed whether the spacecraft was maintaining its range from the ISS and that its acceleration and braking performance remained stable.

Once all the maneuvers had been verified, Dragon re-commenced its approach and demonstrated the second abort scenario, which required the ISS crew to issue a hold command that initiated a period of station-keeping at 220 meters. Once this had been performed, mission controllers verified the Dragon's braking performance was nominal and confirmed the vehicle had stayed within the required range. When all these objectives had been met, Mission Control gave a Go for Close Approach, at which point Dragon fired its engines and closed on the ISS, entering the Keep Out Zone while the ISS crew watched closely to ensure there were no problems during the approach. As the vehicle approached the 30-meter mark, Dragon executed another hold to give the two control centers the opportunity to check the vehicle's status and conduct another Go/No Go poll before allowing Dragon to proceed.

After being approved to continue its approach, Dragon crept towards the ISS before coming to a stop 10 meters from the station. It had reached the Capture Point. When it was verified that Dragon was in the proper position, Free Drift Mode (FDM) was initiated, and Dragon's thrusters were disabled. The rest of Dragon's rendezvous was performed by the station's Canadarm2 robotic arm, under the control of Don Pettit with assistance from Andre Kuipers. First, the Canadarm2 captured Dragon (Figure 6.7) and began a carefully choreographed maneuver to place the vehicle above its intended berthing position at the Earth-facing (nadir) Common Berthing Mechanism (CBM) on the Harmony Module.

Figure 6.7.   Dragon is grappled by the Canadarm2 on May 25, 2012. Credit NASA.

Next, four Ready to Latch Indicators were used to confirm the spacecraft was in the correct position and ready for berthing. Procedures then began to perform first and second stage capture of the spacecraft, at the end of which Dragon was secured in place, forming a hard mate between itself and the ISS. This marked the official start of docked operations. The Canadarm2 then returned to its pre-grapple position to finish the day's work. The rendezvous, up to capture and berthing, had taken about eight hours (Table 6.5).

Table 6.5: Rendezvous Timeline

MET[1]	Event
02/18:51	Rear Height Adjust Burn #2
02/19:30	Read Co-Elliptic Burn #2
02/21:02	Height Adjustment Burn #2
02/21:48	Co-Elliptic Burn #2
02/22:38	Entered ISS Communication Zone
02/23:16	Height Adjustment Burn #3
02/23:32	Mid-Course Correction #1
02/23:50	Mid-Course Correction #2
03/00:02	Co-Elliptic Burn #3
03/00:27	Approach Initiation Burn
03/00:44	Mid-Course Correction #3
03/00:59	Mid-Course Correction #4
03/01:22	R-Bar Acquisition – Range: 350 meters
03/01:22	180° Yaw
03/01:37	Range: 250 meters – station-keeping
03/01:52	Retreat and hold demonstration
03/02:17	Range: 220 meters – hold
03/02:32	Entered 'Keep Out Zone'
03/03:23	Range: 30 meters – hold
03/03:37	Final Approach
03/03:57	Range: 10 meters – Capture Point
03/04:07	Go for Dragon capture
03/04:15	Capture
03/07:36	Berthing

1. Mission Elapsed Time

Flight Day 5 began with a new addition berthed to the ISS. One of the first tasks of the day was to pressurize the vestibule between the Dragon and Harmony Module hatches and check for pressure leaks to make sure the seal between ISS and the vehicle was tight. Once the leak checks were complete, Mission Control gave the 'Go' to open the hatch, at which point crew members began the vestibule outfitting task. This required the installation of ducts and the removal of equipment needed to bolt Dragon in position. Once both hatches were open, the crew conducted air sampling inside Dragon as part of standard ingress operations. Then, Mission Control gave the green flag for Dragon ingress, allowing the crew to begin cargo transfer operations. During the docked phase of the mission, the ISS crew spent about 25 hours conducting cargo operations, which included offloading various items and placing them aboard the station (Table 6.6). Once this task had been completed, the crew loaded cargo on board the Dragon for its return to Earth.

Table 6.6: Dragon C2 Cargo Manifest[*]

ISS CARGO

Food and Crew Provisions: 306 Kilograms:
* 13 bags standard rations: Food, about 117 standard meals, and 45 low sodium meals
* 5 bags low-sodium rations
* Crew clothing
* Pantry items (batteries, etc.)
* SODF and Official Flight Kit

Utilization Payloads: 21 Kilograms
* NanoRacks Module 9 for U.S. National Laboratory: NanoRacks-CubeLabs Module-9 uses a 2-cube unit box for student competition investigations using 15 liquid mixing tube assemblies that function like commercial glow sticks.
* Ice bricks: For cooling and transfer of experiment samples.

Cargo Bags: 123 Kilograms
* Cargo bags: reposition of cargo bags for future flights.

Computers and Supplies: 10 Kilograms
* Laptop, batteries, power supply cables.

Total Cargo Upmass: 460 Kilograms (520 Kilograms including packaging)

RETURN CARGO

Crew Preference Items: 143 Kilograms
* Crew preference items, official flight kit items.

Utilization Payloads: 93 Kilograms
* "Plant Signaling" hardware (16 Experiment Unique Equipment Assemblies): Plant Signaling seeks to understand the molecular mechanisms plants use to sense and respond to changes in their environment.
* Shear History Extensional Rheology Experiment (SHERE) Hardware: SHERE seeks to understand how liquid polymers behave in microgravity by measuring response to straining and stressing.
* Materials Science Research Rack (MSRR) Sample Cartridge Assemblies (Qty 3): MSRR experiments examined various aspects of alloy materials processing in microgravity.
* Other: Supporting research hardware such as Combustion Integrated Rack (CIR) and Active Rack Isolation (ARIS) components, double cold bags, MSG Tapes.

Systems Hardware: 345 Kilograms
* Multifiltration Bed
* Fluids Control and Pump Assembly
* Iodine Compatible Water Containers
* JAXA Multiplexer

EVA Hardware: 39 Kilograms
* EMU hardware and gloves for previous crew members.

Total Cargo Downmass: 620 Kilograms (660 including packaging)

[*]Courtesy NASA

With cargo operations complete, Dragon was closed out, its hatch closed, the ducts removed and control panel assemblies re-installed. Once Harmony's hatch was closed again, the leak check procedure was repeated and the vestibule between the two spacecraft was depressurized to prepare for Dragon's un-berthing (Table 6.7). Once again, the Canadarm2 was used to grapple Dragon before releasing the vehicle. With Dragon flying free, the Canadarm2 maneuvered the vehicle to its release position 10 meters from the ISS. At this point, Dragon was in FDM with all thruster systems inhibited. Dragon's navigation instruments underwent the requisite checkouts to ensure the vehicle was receiving correct navigation data and, with all checks complete, both control centers gave the 'Go' for release. Then Dragon was ungrappled, the Canadarm2 retreated, and Dragon reactivated its thrusters and recovered from FDM. After performing three engine burns to leave the vicinity of the ISS, Mission Control Houston verified that Dragon was on a safe path away from the station.

Table 6.7: Unberthing Timeline

GMT	Event
04:35	Dragon Vestibule Outfitting
04:50	IPCU Deactivation
05:35	Vestibule Depressurization
	Vestibule Leak Checks (65 minutes)
08:05	Unberthing
09:35	Dragon Release

As it moved away from the ISS, Dragon closed its GNC control bay door to protect the instruments during re-entry. Four hours after release, Dragon was at a safe distance from the station and fired its engines before making the deorbit burn, taking it on a trajectory to re-enter Earth's atmosphere. Twenty minutes after the deorbit burn, Dragon hit the entry interface. During re-entry, Dragon's PICA-X heat shield withstood temperatures of 1,600°C. Dragon used its Draco thrusters during this phase to stabilize its position and control its lift to target the landing. About ten minutes before splashdown, at an altitude of 13.7 kilometers, Dragon opened its dual drogue chutes. This then triggered the main chute opening command, which occurred at an altitude of three kilometers. Descending under its main chutes, Dragon slowed to its landing speed of 17 to 20 kilometers per hour and splashed down, landing about 450 kilometers off the California coast.

At the time of Dragon's first flight, SpaceX had also planned to fly a Dragon variant known as DragonLab. Reusable and free-flying, the DragonLab (Table 6.8) would have been capable of carrying pressurized and unpressurized payloads, in common with its CRS sibling, with an advertised up-mass of 6,000 kilograms and a down-mass of 3,000 kilograms. Its subsystems included propulsion, power, thermal and environmental control, avionics, communications, thermal protection, flight software, guidance and navigation systems, and entry, descent, landing and recovery gear. Flights for the DragonLab appeared on the SpaceX manifest until 2017.

Table 6.8: Specifications of the planned DragonLab

Pressure vessel
* 10 m^3 interior pressurized, environmentally controlled, payload volume.
* Onboard environment: 10–46 °C; relative humidity 25~75%; 13.9~14.9 psia air pressure.
Unpressurized sensor bay (recoverable payload)
* 0.1 m^3 unpressurized payload volume.
* Sensor bay hatch opens after orbital insertion to allow sensor access to space and closes upon re-entry.
Unpressurized trunk (non-recoverable)
* 14 m^3 payload volume in 2.3 m trunk, aft of the pressure vessel heat shield; with optional trunk extension to 4.3 m total length, payload volume increases to 34 cubic meters.
* Supports sensors and space apertures up to 3.5 m in diameter.
Power, telemetry and command systems
* Power: twin solar panels provide 1,500 W average, 4,000 W peak, at 28 and 120 VDC.
* Communications: commercial standard RS-422 and military standard 1553 serial I/O, plus Ethernet communications for IP-addressable standard payload service.
* Command uplink: 300 kbps.
* Telemetry/data downlink: 300 Mbit/s standard, fault-tolerant S-band telemetry and video transmitters.

As the cargo variant of Dragon capsules began to ferry cargo to and from the ISS, SpaceX was at work developing the manned version of its spacecraft, initially called DragonRider[3] but more commonly referred to as Crew Dragon. The development of the DragonRider can be traced back to 2006, when Elon Musk announced that SpaceX had built a prototype crew capsule and had tested a 30-man-day life support system. Later, in 2009, Musk suggested the crewed Dragon variant was two to three years away from completion. As is so often the case in the spaceflight arena, this date proved way off the mark. We will examine the development of Crew Dragon in the following chapter, but first it is instructive to assess the significance of its cargo-carrying sibling.

Dragon's May 2012 flight to the ISS reverberated from Cape Canaveral to Capitol Hill and demonstrated that cargo transport to the ISS could be viably outsourced to commercial players. If any skeptics remained after Dragon's return, the first CRS mission for SpaceX that October should have removed any doubt that governments now had someone else they could call on to send their cargo to LEO. Before that, in August 2012, NASA confirmed SpaceX as one of two companies entrusted with returning Americans to space (the other being Boeing), when it awarded the company a $440 million contract to develop the successor to the Shuttle. As the years

---

[3] Going back to the subject of choosing cool names for their spacecraft, the DragonRider moniker may have its origins in the 1997 German children's novel by the same name, written by Cornelia Funke. *Dragon Rider* follows the exploits of a silver dragon named Firedrake, the brownie Sorrel, and Ben, a human boy, in their search for the mythical Himalayan mountain range called the Rim of Heaven.

rolled by, more and more successful Dragon flights accumulated, so it was no surprise when NASA purchased another six cargo flights from SpaceX in January 2016, under a program dubbed CRS2. With the success of Dragon, both in its flight rate and its reusability, SpaceX had once again set the bar for making missions to space cheaper and more accessible. In so doing, the company had created yet another footnote for itself as a formidable force in transforming spaceflight.

7

The Space Taxi Race

"NASA is on a good track to turn over astronaut transportation to commercial operators, and I think ultimately the agency will be successful at doing that."

Comment from Orbital CEO Dave Thompson, despite not winning a CCDev-2 award.

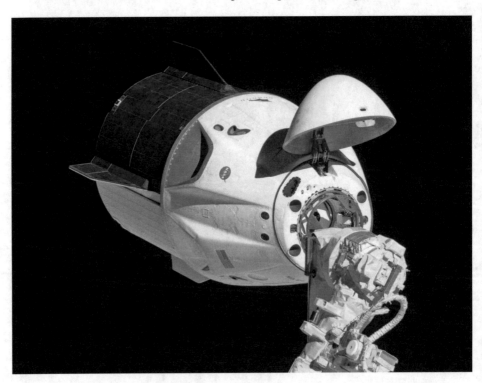

Figure 7.0. Crew Dragon on its first (uncrewed) visit to the International Space Station. Credit NASA.

© Springer Nature Switzerland AG 2022
E. Seedhouse, *SpaceX*, Springer Praxis Books,
https://doi.org/10.1007/978-3-030-99181-4_7

"NASA announces $1.1 billion in support for a trio of spaceships." That was one of the headlines on August 3, 2012, when NASA announced it had committed $1.1 billion over the next 21 months to support spaceship development efforts by a lineup of companies, with the aim of flying American astronauts on American spacecraft within an ambitious deadline of five years. With SpaceX hawking much of the media spotlight following the spectacular success of Dragon's docking with the International Space Station (ISS) earlier that year, it had been easy to forget there were other major players in the race to build space taxis. So, before continuing with the SpaceX story – and Crew Dragon in particular – it is instructive to review some of the other companies competing with SpaceX in developing vehicles capable of flying crews to and from the orbiting outpost. These companies included aerospace juggernaut Boeing, who received $460 million in the Commercial Crew Integrated Capability (CCiCap) outlay, and Sierra Nevada Corporation (SNC), who received $212.5 million.

NASA's CCiCap announcement heralded the next phase of the agency's commercial spaceflight effort. In short, CCiCap called for Boeing, SNC and SpaceX to take their design and testing programs through a series of milestones by May 2014. The goal of the NASA program was to have at least one commercial space taxi ferrying astronauts to and from the ISS by 2017, a deadline that predictably slid to the right. At the time of the announcement, the companies said they were confident they could meet or beat that schedule – provided they continued to receive NASA support. That was good news for the NASA Administrator at the time, Charles Bolden, who was still flying his astronauts on dated Russian hardware at $63 million a seat (SpaceX, by comparison, promised seats for around $22 million, a figure that has since more than doubled).

CCiCap was the third and final phase of NASA's Commercial Crew Program (CCP). In earlier phases, Boeing, SpaceX and SNC had received hundreds of millions of dollars in NASA support. While SpaceX was rapidly upgrading its Dragon capsule to manned capability, Boeing was working on its CST-100 (aka Starliner) and SNC was testing its Dream Chaser space plane, which looked like a scaled down version of the Shuttle. Leading up to the August 2012 decision, NASA and congressional leaders made a deal that called for two commercial partners to receive full funding, with one backup partner receiving half funding. Looking at the numbers, it seemed that SNC drew the short financial straw, since the company's milestones stopped just short of a Critical Design Review (CDR). SpaceX and Boeing could be funded through that phase, but none of the companies were complaining. In a statement, Elon Musk hailed the CCiCap award as "a decisive milestone in human spaceflight" that would set "an exciting course for the next phase of American space exploration." Boeing's statement struck a similar tone, with the company's vice president and general manager of space exploration, John Elbon, announcing; "Today's award demonstrates NASA's confidence in Boeing's approach to provide commercial crew transportation services for the ISS. It is essential for the ISS and the nation that we have adequate funding to move at a rapid pace toward operations, so the United States does not continue its dependence on a single system for human access to the ISS."

The CCiCap funding came partly thanks to the success of SpaceX, because the plan for supporting numerous competitors had been in danger prior to the Dragon cargo flight. Some in Congress, such as NASA Appropriations Chairman Frank Wolf of Virginia, had been pressuring the agency to select a single provider immediately to save money. But Wolf was persuaded to let the competition continue following Dragon's successful flight. This was a good thing for NASA, because the agency did not want to be dependent on a single means of ferrying its astronauts into orbit. They had tried this strategy with the Shuttle and it had not worked out, with the program being grounded for more than five years of the 30 years it was operational (about 17 percent) for accident investigations. Whichever companies NASA ultimately selected for carrying crew, the new vehicles would have to go beyond merely being a Shuttle replacement in one important way: rescue capability. From the time the ISS was first occupied, there has always been a Russian Soyuz capsule berthed to provide a lifeboat capability (Figure 7.1). The reliance on the Soyuz for this service was because the Shuttle did not have the ability to stay docked with the station for more than a week or so (spacecraft serving the lifeboat function must be capable of staying docked with the station for 210 days). So, the new vehicles had to be designed for an orbital life of several months and provide room for up to seven crewmembers. The vehicles also had to be larger than the cramped, claustrophobic confines of the three-passenger Soyuz. The plan was that when Crew Dragon and the CST-100 were finally declared operational the vehicles would be game-changers, not only because they would eliminate NASA's dependence on the Russians (with whom relations can be rocky), but also because these new spacecraft would allow an expansion of ISS crew capacity, which was limited partly by lifeboat capacity (crew size is also partly limited by the constraints of the life support system in scrubbing carbon dioxide from the atmosphere, but that is another story).

Before examining the new crop of commercial vehicles more closely, it is instructive to review the parameters of NASA's CCP. The program was established to invest financial and technical resources to catalyze efforts within the commercial sector to develop and demonstrate safe, reliable, and cost-effective space transportation capabilities. The program managed Commercial Orbital Transportation Services (COTS) partnership agreements with U.S. industry totaling $800 million for commercial cargo transportation demonstrations. It also oversaw Commercial Crew Development (CCDev), a NASA investment funded by $50 million of the American Recovery and Reinvestment Act (ARRA) funds. CCDev-2 (see Appendix III) commercial partners included Blue Origin, Boeing, SNC and, of course, SpaceX. In essence, the CCP was a multiphase strategy that doled out funds to companies to develop solutions for crew transportation to Low Earth Orbit (LEO).

Figure 7.1. Soyuz spacecraft docked with the ISS. Credit NASA.

The August 2012 CCiCap announcement, which was the final-phase development funding under NASA's CCP, established Boeing and SpaceX as the clear frontrunners, with SNC waiting in the wings as the fallback system in case one of the other two faltered. Passed over for a CCiCap award was ATK Aerospace, the long-time supplier of solid rocket motors for the now retired Shuttle. ATK, whose Liberty design had not been funded in previous rounds of the CCP, would have used a self-built solid-fuel core stage and the first stage of Europe's Ariane 5 rocket as an upper stage. Meanwhile, Boeing and SpaceX were focused on meeting their NASA-approved milestones in the 21-month CCiCap performance period. If they managed that, which they did, their designs would undergo a CDR. If the CDRs went well, which they did, construction could begin. As for Blue Origin, the secretive space startup bankrolled by Amazon.com founder Jeff Bezos did not submit a CCiCap proposal, although the company had been involved in NASA's CCP since the first round of funding was awarded in 2010. The company, based in Kent, Washington, had received $25.7 million in NASA CCDev-2 funding, some of which it had put toward a crew escape system for its New Shepherd Vertical-Takeoff, Vertical-Landing (VTVL) suborbital vehicle.

In its assessment of the proposals, NASA showed a clear preference for complete transportation systems rather than merely subsystems, such as those proposed by companies like Paragon Space Development Corporation which won one of the five first-round CCDev awards to work on a life support system. Other companies were cut during the assessment for having major weaknesses or for simply failing to follow instructions. The eight finalists – ATK, Blue Origin, Boeing, Excalibur Almaz, OSC, SNC, SpaceX, and United Launch Alliance (ULA) – were re-evaluated based on technical and business criteria, although these evaluations served as indicators only and did not form the sole basis of the final decision. What NASA was looking for was a diverse range of technical approaches, a strong business approach and spacecraft development versus launch vehicle development. This last requirement made sense, since the U.S. already had plenty of launch vehicle development expertise and experience but comparatively little experience developing crew-carrying spacecraft. When it came to selecting the winning three, the Boeing and SpaceX proposals stood out from the rest thanks to their high ratings in technical and business factors. It also helped that the two companies were developing capsules, whereas SNC was developing a winged spacecraft. What follows is a brief examination of the designs of the three contenders, starting with Boeing (see sidebar: *Boeing*).

BOEING

Profile

Spacecraft: CST-100
Type: Capsule with service module
Crew Capacity: 7
Launch Vehicle: Atlas V (United Launch Alliance)
CCiCAP Funding: $460 million
CCiCAP Term: 21 months
Previous CCDev Funding (including optional milestones): $130.9 million (Boeing), $6.7 million (ULA)
Total CCDev and CCiCAP Funding (if all milestones met): $590.6 million (Boeing), $6.7 million (ULA)

Propulsion
Crew Module: 12 reaction control system (RCS) thrusters; Service Module: 28 RCS thrusters; 20 orbital maneuvering and attitude control (OMAC) thrusters; 4 launch abort engines

Dimensions
Height: 5 meters. Diameter: 4.6 meters

Ascent abort landing requirements
Wave height below 4 meters. Surface winds below 27 knots. No thunderstorms or lightning within abort landing area

Landing constraints
Average near-surface (33.5 meters) wind speed not to exceed 10.3m/sec (11.8m/sec in a contingency). Temperature to be no lower than -9.4°C. Cloud ceiling to be no lower than 305 meters with 1.9-kilometer visibility.

Figure 7.2. Boeing's Starliner. Credit NASA.

Rather than use heritage technology from the lifting body program, Boeing advanced plans for a capsule-based spaceship: the CST-100 capsule (Figure 7.2). In common with the other competitors in the space taxi race, the CST-100's primary mission would be to transport crew to the ISS, and possibly to private space stations such as Bigelow's Aerospace Commercial Space Station or Axiom's space station (Figure 7.3). At first glance, the gumdrop-shaped capsule looks similar to the Apollo and Orion, the latter a spacecraft being built for NASA by Lockheed Martin to be flown on the Space Launch System (SLS).

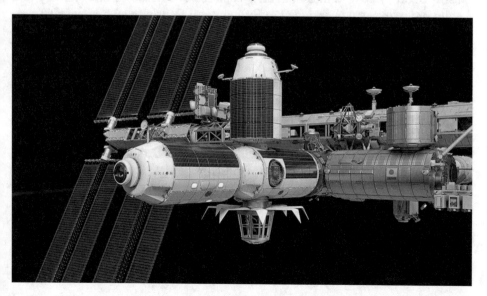

Figure 7.3. Axiom's space station, shown here attached to the ISS. Axiom's plan is to separate its station to become a free flying platform. Credit Axiom.

When complete, the CST-100 (incidentally, the number '100' stands for 100 km, the height of the Kármán line which defines the boundary of space) will be larger than the Apollo command module but smaller than the Orion. Capable of ferrying crews of up to seven thanks to a generous habitable interior and the reduced weight of equipment needed to support an exclusively LEO configuration, the CST-100 is designed to remain on orbit for up to seven months and to be re-used for up to ten missions. While the launch vehicle for the CST-100 will be the Atlas V, the spacecraft is compatible with other launch vehicles, including the Delta IV. In common with SNC, Boeing was also the recipient of earlier NASA funding. In the first phase of the CCDev program the company was awarded $18 million for preliminary development of the spacecraft, which was followed by another $93 million in the second phase for further development. Although an industry juggernaut, Boeing was still reliant on external funding for development of the CST-100. In fact, in July 2010 the company stated the capsule would only become operational by 2015 with sufficient near-term approvals and funding. Boeing also indicated they would only proceed with development of the CST-100 if NASA implemented the commercial crew transport initiative that had been announced by the Obama Administration in its Fiscal Year 2011 (FY11) budget request. Boeing's business case was not only reliant on continued NASA funding, but also on the existence of a second destination, hence the partnership with Bigelow.

While the CST-100 is sometimes confused with its (very) big budget cousin Orion, the Boeing vehicle has no Orion heritage, its design drawing mostly upon Boeing's experience with the Apollo, Shuttle and ISS programs, and the Orbital Express project from the Department of Defense. Measuring 4.56 meters across at its widest point and standing 5.03 meters high, the CST-100 sports a four-engine 'pusher abort system' rather than an escape tower as used in the Mercury and Apollo programs, and is designed to land on terra firma although it can also support a water landing in the event of an abort. For docking with the ISS the vehicle will use the Androgynous Peripheral Attach System (APAS), while for re-entry the Boeing Lightweight Ablator (BLA) heat-shield will protect the crew. Work on CST-100 design, manufacture, testing and evaluation proceeded rapidly thanks to Boeing pulling in proven technology. For example, the CST-100 utilizes technology developed to build the Apollo-era heat shield and the Shuttle's thermal protection system, as well as autonomous rendezvous and docking gear developed on the Pentagon's experimental satellite-refueling Orbital Express mission. Given that Boeing was working on a fixed-price development, using mature designs, *and* drawing upon flight-proven hardware, it not only made sense financially, but also gave them an edge on the competition. Surprisingly, despite all the many advantages Boeing had over its nearest competitor, it was still not enough to beat SpaceX. Not even close.

Figure 7.4. CST-100 undergoing a drop test at Langley. Credit NASA.

Some of the more notable tests performed by Boeing in its development of the CST-100 included drop tests (Figure 7.4) to validate the design of the air bag cushioning system and the capsule's parachute system. The air bags, which are deployed by filling the bags with a mixture of compressed nitrogen and oxygen, are located underneath the CST-100's heat shield, with the latter designed to be separated from the capsule while under canopy at about 1,500 m altitude. In September 2011, Boeing conducted drop tests in the Mojave Desert in California, at ground speeds of between 16 and 48 km/h to simulate cross wind landing conditions. Bigelow Aerospace, one of Boeing's partners in the CST-100, built the mobile test rig and conducted the tests. The drop tests were followed by parachute tests in April 2012, when Boeing dropped a CST-100 mock-up over the Nevada desert at Delamar Dry Lake, successfully testing the spacecraft's three main landing parachutes from 11,000 feet. Bigelow Aerospace President, Robert Bigelow, was impressed, noting in a press statement: "If astronauts had been in the capsule during these drop tests, they would have enjoyed a safe, smooth ride… further proof that the commercial crew initiative represents the most expeditious, safest, and affordable means of getting America flying in space again." Since then, Boeing has followed the Commercial Crew milestone checklist (Table 7.1), which it completed in 2014.

Table 7.1: Boeing Commercial Crew Base Milestones.

No.	Description	Original Date	Amount
1	**Integrated System Review.** Boeing shall conduct an Integrated Systems Review (ISR) which establishes and demonstrates a baseline design of the Commercial Crew Transportation System (CCTS) integrated vehicle and operations that meets system requirements.	August 2012	$50 Million
2	**Production Design Review.** Boeing shall conduct a Production Design Review which establishes the baseline plan, equipment, and infrastructure for performing the manufacture, assembly, and acceptance testing of the CST-100 spacecraft.	October 2012	$51.7 Million
3	**Safety Review Board.** Boeing shall prepare and conduct a Phase 1 Safety Review of the CCTS Preliminary Design Review (PDR) level requirements, vehicle architecture and design, and associated safety products, to assess conformance with NASA Crew Transportation System certification process (PDR-level products).	November 2012	$25.2 Million
4	**Software Integrated Engineering Release 2.0.** Boeing shall demonstrate the software release [REDACTED] closed loop with Guidance, Navigation & Control (GN &C) for the flight ascent phase.	January 2013	$20.4 Million
5	**Landing & Recovery / Ground Communication Design Review.** Boeing shall conduct a Landing & Recovery / Ground Communication Design Review which establishes the baseline plan, for equipment, and infrastructure, for conducting CST-100 spacecraft flight operations, fulfilling both ground communications and landing and recovery operations.	January 2013	$28.8 Million
6	**Launch Vehicle Adapter (LVA) Preliminary Design Review (PDR).** The LVA PDR demonstrates that the preliminary design meets requirements with acceptable risk and within the cost and schedule constraints, and establishes the basis for proceeding with detailed design.	February 2013	$45.5 Million
7	**Integrated Stack Buffet Wind Tunnel Test.** Boeing shall develop a test matrix, fabricate the necessary test models, and perform an integrated launch vehicle force and moment wind tunnel test, to validate predictions on integrated Crew Module (CM)/Service Module (SM)/Launch Vehicle (LV) stack for ascent.	April 2013	$37.8 Million

Table 7.1:   (continued)

8	**Dual Engine Centaur (DEC) Liquid Oxygen Duct Development Test.** Boeing shall complete a Dual Engine Centaur Liquid Oxygen Duct Development Test.	May 2013	$21.5 Million
9	**Orbital Maneuvering and Attitude Control (OMAC) Engine Development Test.** Boeing shall complete the OMAC Engine development test to support component, subsystem and CST-100 vehicle level development.	July 2013	$50.2 Million
10	**Spacecraft Primary Structures Critical Design Review (CDR).** A Spacecraft Primary Structures CDR confirms that the requirements, detailed designs, and plans for test and evaluation form a satisfactory basis for fabrication, assembly and structural testing.	October 2013	$8.6 Million
11	**Service Module Propulsion System Critical Design Review.** Boeing shall perform a Service Module (SM) Propulsion System Critical Design Review (CDR) after major SM Propulsion components have completed their individual CDR.	November 2013	$7.5 Million
12	**Mission Control Center Interface Demonstration Test.** The Mission Control Center (MCC) Interface Demonstration Test demonstrates the linkage between the MCC and the Boeing Avionics Software Integration Facility, which is a precursor to integrated simulation capability for flight operations training.	September 2013	$7.9 Million
13	**Launch Vehicle Adapter Critical Design Review.** Boeing shall complete a Launch Vehicle Adapter (LVA) Critical Design Review (CDR). CDR confirms that the requirements, detailed designs, and plans for test and evaluation form a satisfactory basis for production and integration.	September 2013	$13.5 Million
14	**Emergency Detection System (EDS) Standalone Testing.** Boeing shall complete the Initial EDS Testing – Launch Vehicle Stand-alone.	October 2013	$13.8 Million
15	**Certification Plan.** Boeing shall complete a review of the CCTS Certification Plan. Review with NASA the Boeing Certification Plan which defines our strategy leading to a crewed flight test.	November 2013	$5.8 Million
16	**Avionics Software Integration Lab (ASIL) Multi-String Demonstration Test.** Boeing shall demonstrate the flight software closed loop with GN&C for the flight ascent phase.	December 2013	$24.9 Million

Table 7.1: (continued)

17	**Pilot-in-the-loop Demonstration.** Boeing shall demonstrate key hardware/software interfaces for Manual Flight Control meets requirements, including operational scenarios and failure modes.	February 2014	$13.9 Million
18	**Software Critical Design Review.** Boeing shall conduct a Spacecraft Software CDR. CDR confirms that the requirements, detailed designs, and plans for test and evaluation form a satisfactory basis for flight software development, verification, and delivery.	March 2014	$15.1 Million
19	**Critical Design Review (CDR) Board.** Boeing shall establish and demonstrate a critical baseline design of the CCTS that meets system requirements. CDR confirms that the requirements, detailed designs, and plans for test and evaluation form a satisfactory basis for production and integration.	April 2014	$17.9 Million
21A	**Boeing Spacecraft Safety Review.** Boeing shall prepare and conduct a Phase 2 Safety Review of the Commercial Crew Transportation System (CCTS) Spacecraft Critical Design Review (CDR) level requirements, system architecture and design, and associated safety products to assess conformance with Commercial Crew Transportation System certification process (CDR-level products).	July 2014	$20 Million

By the time Boeing had completed its milestones, SpaceX had attained 13 of its own 18 targets, but NASA granted the company an extension to complete the work. The same applied to SNC, which had also fallen behind Boeing. After all that SpaceX had accomplished, and the speed at which they had accomplished the development and manufacture of their launch vehicles, for the company to be trailing Boeing was a surprise. But developing Crew Dragon was not the only problem facing Musk. SpaceX had planned for 24 Falcon 9 flights in 2013 but had only flown seven. There was also talk of anomalies that had occurred during some CRS flights, which included helium leaks, loss of spacecraft control and thruster issues. All these problems were of concern because Crew Dragon was based heavily on the design of Cargo Dragon.

Shortly after Boeing had completed its milestones, NASA announced Boeing and SpaceX as the two companies that would be funded to develop spacecraft to ferry crews to the ISS. Boeing won $4.2 billion to complete its Starliner by 2017 and SpaceX won $2.6 billion to complete and certify Crew Dragon. The contracts included at least one crewed flight test with at least one NASA astronaut on board. For Boeing, the contract stipulated that it should perform at least two and as many as six missions to the ISS. It is worth mentioning that, in 2014, many in the industry, including NASA's Associate Administrator for Human Explorations and

Operations, William Gerstenmaier, considered Boeing's spacecraft a stronger candidate than Crew Dragon. The reason this is noteworthy is because, at the time of writing, the Starliner has already lost the space taxi race and is still a long, *long* way from flying a crew. Incidentally, the contract also included an agreement permitting Boeing to sell seats for spaceflight participants at a price that would be competitive with Roscosmos. But with its seat price set at more than $90 million, the Starliner is evenly matched with the seat price charged by the Russian company. Not that any of that matters, because the Soyuz is flying and the Starliner, as of the end of 2021, remains firmly grounded and not flying anywhere.

In September 2015, Boeing announced the official name of what would soon become a troubled spacecraft. It would be known as the CST-100 Starliner. Two months later, NASA announced the agency had dropped Boeing from being considered to fly CRS missions to the ISS. Then, in May 2016, Boeing announced the first of what would be many slips to its schedule, saying the first Starliner launch would take place in early 2018 and not in 2017 as planned. This slip was closely followed by a second one when Boeing delayed the program by *another* six months, saying they hoped to fly astronauts to the ISS in December 2018. NASA did not share Boeing's optimism and suggested the first crewed Boeing flight would take place in 2019. Or 2020. If there were no further delays.

Unfortunately, there were. One reason for the delays was weight loss, with the spacecraft having gained too much weight while being developed. Boeing now needed to trim that excess mass to meet the requirements of the launch and ascent on the Atlas V rocket. Now you may be wondering how a company such as Boeing could have allowed a weight issue to become a problem, but the engineers had simply exceeded the margins for mass growth allowances, which meant the spacecraft had outgrown the lift capability of the Atlas V rocket. Of course, one option would have been to add a third strap-on booster to the Atlas V to compensate (this option was considered), but ultimately Boeing managed to trim the extra weight, only to be confronted by a problem concerning aerodynamic loads (Figure 7.5). This issue came to light after wind-tunnel testing of the spacecraft's qualification unit. The problem was that the capsule's outer shape had to be changed because the wind tunnel testing had revealed that higher than predicted aerodynamic loads would be placed on the vehicle during a launch. Boeing had to go back to the drawing board again. At the same time Boeing was dealing with the mass and aerodynamic load issues, engineers were also going about the time-consuming business of qualifying each component on the spacecraft. An example of just how time-consuming this process is can be gleaned from the avionics boxes. The CST-100 has 200 of these, each of which must undergo a unique level of qualification. While the qualification is conducted by the vendor, all 200 boxes must be integrated into the avionics systems and subsystems so the whole kit and caboodle functions correctly and interfaces functionally with the other systems, such as the electrical system or the life support system. For an aerospace juggernaut like Boeing, all this should have been business as usual, but the company very soon lost its way.

Figure 7.5.   The CST-100 in its 2020 configuration after excess weight had been stripped and the shape modified to deal with aerodynamic loads. Credit NASA.

Having dealt with all these issues, Boeing announced that the first (unmanned) flight test would take place in the spring of 2019, with a manned test flight scheduled for the summer that same year, depending on the results of the unmanned fight. Those plans quickly slid to the right in July 2018 when faulty valves led to a hypergolic propellant leak, prompting a mission reset to August 2019. Before that could happen however, Boeing had to conduct a pad abort test, which was delayed until November 2019. The test started promisingly, but quickly went sideways when one of the three parachutes failed to deploy. The capsule landed safely, and Boeing said the incident would not affect the development. Eventually, the time came for the first orbital flight test, which took place on December 19, 2019. The launch was successful (Figure 7.6), but shortly after reaching orbit an 11-hour offset in the mission clock resulted in the onboard computer incorrectly computing that the spacecraft was in an orbital insertion burn. This resulted in the spacecraft burning more fuel than planned, which in turn resulted in the planned docking with the ISS being cancelled. Incidentally, if the software issue had not been detected and corrected, the spacecraft could have been destroyed.

Figure 7.6. An Atlas V rocket carrying Boeing's hexed CST-100 Starliner launches from Space Launch Complex 41, Friday, December 20, 2019, at Cape Canaveral Air Force Station in Florida. Credit NASA.

The spacecraft landed safely, but it was hardly a successful mission. A joint NASA-Boeing investigation noted the following three concerns:

1. An error with the Mission Elapsed Timer (MET), which incorrectly polled time from the Atlas V booster nearly 11 hours prior to launch.
2. A software issue within the Service Module (SM) Disposal Sequence, which incorrectly translated the SM Disposal Sequence into the SM Integrated Propulsion Controller (IPC).
3. An Intermittent Space-to-Ground (S/G) forward link issue, which impeded the Flight Control team's ability to command and control the vehicle.[1]

The outcome of the investigation was troubling because the team found that despite multiple safeguards, the critical software defects that caused the problem had not been identified. Significantly, these were anomalies that could have caused loss of the vehicle. Ultimately, the cause of the debacle was a failure of Boeing's

---

[1] Taken from https://blogs.nasa.gov/commercialcrew/2020/02/07/nasa-shares-initial-findings-from-boeing-Starliner-orbital-flight-test-investigation/2/

software quality control procedures. So, Boeing went to work to correct the issue and began planning for the second orbital flight test. Since the first flight test had gone awry, the second flight test would be unmanned and Boeing said it would fund the costs of the second flight test, expected to be around $410 million. Due to the software issues, the second test was planned for March 2021 with a crewed flight planned for the summer of 2021.

> *"This is, obviously, a disappointing day. This team is going to go figure this out and we will go fly when we are ready. You might be hearing us sound a little tired, because teams have been working really hard, but we're not frustrated. I'm a little sad, because I wanted to go fly, but I'm also very proud of the team."*

Kathy Lueders, NASA Associate Administrator for Human Exploration and Operations, in a call with reporters, August 13, 2021.

After some delays, the second orbital flight was set to launch in August 2021, but the hex bedeviling Boeing continued. First, the Atlas V, which had been rolled out to the pad on July 29, was immediately rolled back because of drama on the space station. The long overdue Nauka module had docked with the space station earlier that day but had malfunctioned (a thruster misfiring), causing the station to lose attitude. The Atlas V was rolled out to the pad again on August 2, with a plan to launch the next day. The August 3 launch attempt was scrubbed due to propulsion problems with the Starliner, so the launch clock was reset for 24 hours later. The August 4 attempt was also scrubbed because valve positions in the propulsion system were incorrect (causing nitrogen tetroxide to permeate the Teflon seals in the valves), prompting another roll back of the Atlas V. Then, on August 13, following days of troubleshooting, Boeing took the decision to return the spacecraft to the processing facility, a decision that pushed the flight into 2022. At the time of writing, the Boeing saga continues, with some suggesting that the Dream Chaser (Figure 7.7) could be flying with crew before the Starliner. That would really rub salt into Boeing's wounds.

> *"I believe firmly that when the company celebrates its second hundred years, there will be a division of Boeing that's building commercial space vehicles, that will be of that magnitude, of that size."*

John Elbon, who heads up Boeing's space exploration division.

Figure 7.7. Sierra Nevada Corporation's Dream Chaser spacecraft. The astronaut favorite. Credit Sierra Nevada Corporation.

SIERRA NEVADA CORPORATION

Profile

Location: Louisville, Colorado
Spacecraft: Dream Chaser
Type: Lifting body
Crew Capacity: 7
Launch Vehicle: Atlas V (United Launch Alliance)
CCiCAP Funding (if all milestones met): $212.5 million
CCiCAP Term: 21 months
Previous CCDev Funding: $125.6 million (Sierra Nevada), $6.7 million (ULA)
Total CCDev and CCiCAP Funding (if all milestones met): $338.1 million (Sierra Nevada), $6.7 million (ULA)

SNC's private space plane (see sidebar: *Sierra Nevada Corporation*) aimed to pick up where the Shuttle left off. At least, that was the impression when viewing a vehicle that could easily have passed for a miniature Shuttle (Figure 7.7). But while the spacecraft's design certainly took cues from the past, the company hoped its winged spaceplane – Dream Chaser – would redefine commercial spaceflight in much the same way the Shuttle changed manned spaceflight in the 1980s. Headquartered in Louisville, Colorado, SNC's Space Systems division designs and manufactures spacecraft, rocket motors, and spacecraft subsystems for the U.S. Government and commercial customers. While not in the same category as Boeing, SNC Space Systems has more than 400 successful space missions under its belt, has delivered more than 4,000 systems, subsystems and components, and has concluded over 70 programs for NASA.

The Dream Chaser was designed as a crewed suborbital / orbital Vertical-Takeoff, Horizontal-Landing (VTHL) lifting-body spaceplane that could carry up to seven crew to and from LEO. It was originally planned in 2004 to be a suborbital vehicle modeled after the X-34, but the design was revised in 2005 and was ultimately based on NASA's HL-20 lifting body concept. Intended to launch vertically on a man-rated Atlas V 402 rocket and land horizontally on conventional runways, the primary mission of the reusable composite Dream Chaser, as with Boeing's Starliner, was to provide NASA with a safe and reliable transportation service for crew and cargo to and from the ISS. Potential missions included delivering crew and cargo to other orbiting facilities such as Bigelow's habitats, functioning as a short term independent orbiting laboratory for other government agencies, and even orbital space tourism.

In common with the Shuttle, the Dream Chaser would have returned to Earth by gliding, but while the Shuttle was restricted to landing on two NASA runways, the Dream Chaser could have landed on any airport runway capable of handling commercial air traffic. The vehicle design featured a launch escape system powered by a hybrid-motor pusher system, to eject from a failing Atlas V and fly a piloted return-to-landing-site maneuver with a 2G load on the crew. For maneuvering in space, the vehicle would have used its ethanol-fueled Reaction Control System (RCS). The advantage of using ethanol, which is not an explosively volatile material, is that it would have allowed the Dream Chaser to be handled immediately after landing, unlike the Shuttle which required a small army to make it safe before being readied for its next flight. Another improvement on the Shuttle was Dream Chaser's Thermal Protection System (TPS), an ablative tile created by NASA's Ames Research Center (ARC), that would have been replaced as a large group rather than the laborious and time intensive tile-by-tile process that was the case with the Shuttle.

Publicly announced on September 20, 2004 as a candidate for NASA's Vision for Space Exploration (VSE) and later for the agency's Commercial Orbital Transportation Services Program (COTS), the Dream Chaser was not selected under Phase 1 of the COTS Program. This setback resulted in founder Jim Benson stepping down as Chairman of SpaceDev and starting Benson Space Company to pursue development of the vehicle. In April 2007, SpaceDev announced it had partnered with United Launch Alliance (ULA) to pursue the possibility of utilizing the Atlas V as Dream Chaser's launch vehicle. This partnership was followed by SpaceDev's acquisition by SNC in December 2008. Then, on February 1, 2010, SNC was awarded $20 million in seed money under CCDev Phase 1 for the development of Dream Chaser. Using this funding, SNC completed four planned milestones, including program implementation plans, manufacturing readiness capability, hybrid rocket test firings, and preliminary structure design. Additional tests included a drop test of a 15 percent scaled version (at the NASA Dryden Flight Research Center) to test flight stability and collect aerodynamic data for flight control surfaces.

Later that year, for the CCDev Phase 2 solicitation in October 2010, SNC proposed extensions of Dream Chaser technology. Following the CCDev Phase 2 solicitation, SNC announced it had achieved two critical milestones for the program. The first consisted of three successful test firings of a single hybrid rocket motor in one day and the second was completion of the primary tooling necessary to build the composite structure of the vehicle. Six months after achieving these milestones, on April 18, 2011, NASA awarded another $80 million in CCDev funding to SNC. Following this funding, SNC completed nearly a dozen further milestones, including testing of the airfoil fin shape, integrated flight software and hardware, landing gear, and a full scale captive carry flight test. These milestones were followed by a System Requirements Review, a new cockpit simulator, finalizing the tip fin airfoil design, and a Vehicle Avionics Integration Laboratory (VAIL) designed to test Dream Chaser computers and electronics in simulated space missions.

Then, in June 2011, SNC signed a Space Act Agreement (SAA) with NASA, which was followed by ULA announcing that the Atlas V would be used to launch Dream Chaser. By February of 2012, SNC had completed the assembly and delivery of the primary structure of the first Dream Chaser flight test vehicle. With this achievement, SNC had completed all 11 CCDev milestones on time and on budget. Two months later, the company announced the successful completion of wind tunnel testing of a scale model of the Dream Chaser vehicle and, on May 29, 2012, a captive carry test was conducted near the Rocky Mountain Metropolitan Airport, to determine the spacecraft's aerodynamic properties. This test was followed by buffet tests on Dream Chaser and the Atlas V stack, and aerodynamic and aerothermal analysis by the Langley/SNC team. On July 11, 2012, SNC announced they had successfully completed testing of the Dream Chaser's nose landing gear during simulated approach and landing tests, as well as the impact of future orbital flights.

Yet more funding followed on August 3, 2012, when NASA announced the award of $212.5 million to SNC to continue work on the Dream Chaser under the CCiCap Program. The 21-month contract began in August of 2012. With this funding, the Dream Chaser seemed to be in a good position to achieve operational commercial human flight capability as early as 2016. Reaching this point had only been possible because of SNC having integrated the efforts of a powerful team of aerospace companies, academic institutions and NASA Centers to advance the development of the Dream Chaser and the mission, ground, and crew systems. After completing a full system PDR and the first captive carry flight, SNC looked forward to the approach and landing test scheduled for later in 2012, a test mirroring the first flight test of the Shuttle. SNC had ten CCiCap milestones to meet during the 21-month base period (Table 7.2). The company received what amounted to half an award, a sum that was significantly less than the ones provided to Boeing and SpaceX. The company also had 31 optional milestones for which it would have received funding if NASA had money available. These milestones were redacted from the SAA for competitive reasons.

Table 7.2: SNC Commercial Crew Milestones.

No.	Description	Original Date	Amount
1	**Program Implementation Plan Review**. This was an initial meeting to describe the plan for implementing the CCiCap, to include management planning for achieving CDR; Design, Development, Testing, and Evaluation activities; risk management to include mitigation plans, and certification activities planned during the CCiCap Base Period.	August 2012	$30 Million
2	**The Integrated System Baseline Review (ISBR)** demonstrated the maturity of the baseline CTS integrated vehicle and operations design of the Dream Chaser Space System, consisting of the spacecraft, the Atlas launch vehicle, Mission Systems, and Ground Systems support.	October 2012	$45 Million
3	**Integrated System Safety Analysis Review #1** demonstrated that the systems safety analysis of the Dream Chaser Space System had been advanced to a preliminary maturity level, incorporating changes resulting from the PDR.	January 2013	$20 Million
4	**Engineering Test Article Flight Testing** reduces risk due to aerodynamic uncertainties in the subsonic approach and landing phases and matures the Dream Chaser aerodynamic database. Up to five Engineering Test Article free flight tests were planned to be completed to characterize the aerodynamics and controllability of the Dream Chaser outer mold line configuration during the subsonic approach and landing phase.	April 2013	$15 Million
5	**SNC Investment Financing #1.** This funding represented SNC's commitment for significant investment financing.	July 2013	$12.5 Million
6	**Integrated System Safety Analysis Review #2** demonstrated the systems safety analysis of the Dream Chaser Space System (DCSS).	October 2013	$20 Million
7	**Certification Plan Review** defined the top-level strategy for certification of the DCSS that met the objectives for the ISS Design Reference Mission.	November 2013	$25 Million
8	**Wind Tunnel Testing** would have reduced the risk on the vehicle by helping determine pre-CDR required updates to the OML or control surface geometry if required.	February 2014	$20 Million
9	**Risk Reduction and TRL Advancement Testing.** The purpose of these tests was to significantly mature all Dream Chaser systems to or beyond a CDR level.	May 2014	$17 Million
9a	**Main Propulsion and RCS Risk Reduction and TRL Advancement Testing.** The purpose of these tests was to significantly mature the Dream Chaser Main Propulsion System and Reaction Control System to or beyond a CDR level. Risk reduction and Technology Readiness Level improvement tests were planned to be completed for these systems.	May 2014	$8 Million
		TOTAL: $212.5 Million	

In NASA's source-selection document, the Dream Chaser was critiqued for posing significant risks due to design complexity. While the document acknowledged that a winged vehicle offered advantages to customers in terms of easier landings, lower Gs and greater cross-range from orbit, NASA also pointed out that aborts and thermal protection issues were more difficult. It was due to these technology hurdles that the agency cut back on the milestones SNC had proposed, commensurate with the lower funding level. The good news was that SNC decided to go ahead with its bid to fly NASA astronauts, with plans to conduct autonomous drop tests and landing with a full-scale engineering article. Ultimately, SNC believed, the capabilities of Dream Chaser would outweigh the risks stated in the source-selection document.

The company's optimism seemed to be well-placed. As August 2012 rolled around, the company had completed CCiCap Milestone 1 and followed that up by completing the Integrated System Baseline Review (CCiCap Milestone 2) two months later, a check in the box that demonstrated Dream Chaser's maturity. Captive carry tests were announced and conducted in 2013, the year in which former Shuttle commander Lee Archambault joined SNC to work as test pilot for Dream Chaser. Dream Chaser's prospects were looking good. Then, in September 2014, NASA announced that Dream Chaser had not been selected in the latest round of CCiCap, prompting SNC to file its protest with the US Government Accountability Office (GAO). SNC lost that dispute, but still decided to pursue their effort of sending crews to orbit. They did this by transforming Dream Chaser into the Dream Chaser Cargo System (DCCS), a vehicle that will begin flying cargo to the ISS sometime in 2022. But SNC is also developing the spacecraft so that it can carry crew, which means there is a possibility it can beat Boeing to orbit.

SpaceX

Profile

Location: Hawthorne, California
Spacecraft: Crew Dragon
Type: Capsule
Crew Capacity: 7
Launch Vehicle: Falcon 9
CCiCAP Funding (if all milestones met): $440 million
CCiCAP Term: 21 months
Previous CCDev Funding: $125.6 million
Total CCDev and CCiCAP Funding (if all milestones met): $911 million
(see Figure 7.8 for a list of the milestones)

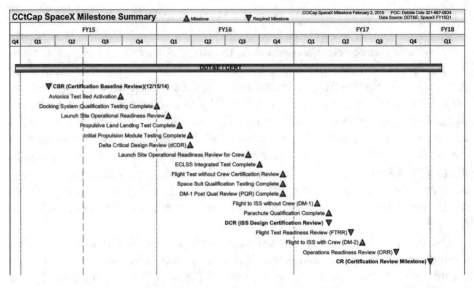

Figure 7.8. SpaceX milestone summary.

Figure 7.9. The Demo-2 Crew Dragon prior to acoustic testing as part of prelaunch processing at Cape Canaveral Air Force Station in February 2020. The spacecraft, which is partially reusable, can carry 3,307 kilograms and up to seven astronauts. It features touchscreen control panels, a toilet and a docking hatch. Eight SuperDraco engines are side mounted around the capsule, with each engine capable of generating 71 kN thrust. Docking is executed autonomously using the Canadarm2. Incidentally, the spacecraft can be operated in a vacuum. A PICA-X heatshield protects the vehicle during re-entry, while a moveable ballast sled permits attitude control during its return to Earth. Credit SpaceX.

"This is a unique moment where all of America can take a moment and look at our country do something stunning again, and that is launch American astronauts on American rockets from American soil. We're transforming how we do spaceflight in general."

Jim Bridenstine, NASA's Administrator, speaking ahead of the Demo-2 launch

After nearly 64 days (63 days, 23 hours and 25 minutes) in space, NASA astronauts Bob Behnken and Doug Hurley (Figure 7.10) landed back on Earth following a flight that many pundits labeled a landmark mission. Which it was in many ways. Firstly, Hurley and Behnken not only became the first crew to fly the Crew Dragon (Figure 7.9 and sidebar: *SpaceX*), but their mission was the first crewed flight from home soil in almost nine years (no more writing checks to the Russians to use their Soyuz space taxi). The mission was also heralded as representing a much larger milestone in spaceflight history, namely proving that a private spacecraft *could* ferry astronauts to space and back.

Figure 7.10. NASA's crew choice for the SpaceX Demo-2 mission was marked by some unusual symmetry. Bob Behnken and Doug Hurley both graduated from the same astronaut class (the 2000 selection). Each is an engineer by profession, and each has experience flying military aircraft. If that was not enough symmetry for you, consider the fact that not only had each astronaut flown in space three times, but each had married a fellow astronaut (Behnken is married to Megan McArthur and Hurley is married to Karen Nyberg). They also both shared the same ambition – to fly a new spacecraft. Credit NASA.

Of course, this was more than a SpaceX mission, because without NASA galvanizing the commercial spaceflight industry with its CCDev program, the Demo-2 mission probably would not have happened and the NASA Administrator would still be writing checks to the Russians. The mission was also noteworthy because industry juggernaut Boeing had failed to deliver, thanks to egregious software glitches. SpaceX won the space taxi race, comfortably, and in winning they opened the door for a new kind of space race. Until the Demo-2 mission, space exploration had been a decidedly state-run political affair. But with NASA having outsourced elements of space travel to companies such as SpaceX, it has created a catalyst for all sorts of opportunities (take the Inspiration4 mission – discussed later in this chapter – as an example).

So, NASA had a big role to play, but an equally big role was played by the disruptive thinking of Elon Musk, who has shown time and time again that space exploration is no longer the sole purview of government entities. Innovative ideas such as those pioneered by the Mars Messiah and his Musketeers have opened the door to unlimited possibilities, as we shall see in the following chapter. But before we do that, let us look at Crew Dragon.

The first time the public got a glimpse of Crew Dragon was in May 2014 when Musk unveiled the spacecraft design. Like the Apollo capsule that carried astronauts to the Moon, Crew Dragon (Table 7.3) is a capsule spacecraft attached to a trunk. The trunk features solar panels, radiators, cargo space, and fins for stability during an abort. The spacecraft also sports 16 Draco thrusters that enable on-orbit maneuvering. On top of the vehicle is a Launch Escape System (LES) comprising eight SuperDraco engines. The interior of the spacecraft (Figure 7.11) features three touchscreens, comfy couches (four is the usual number for most missions, but Crew Dragon can carry up to seven astronauts) that offer plenty of legroom, and in the ceiling, the feature that no one talks about – the toilet. Not much is known about the toilet, but we do know a fair amount about other life support subsystems, so it is instructive to look at these.

Unlike the ISS which features six life support subsystems, Crew Dragon's life support system is divided into two groups, one being air revitalization and the other ox nitrox. The functions of the air revitalization system are to clean, decontaminate and condition the air, while the function of the ox nitrox is to maintain pressure and release breathing gases. Most of the Environmental Control Life Support System (ECLSS in NASA and SpaceX parlance) subsystems and assemblies are located under the floor of the vehicle (Figure 7.12).

The Pressure Control subsystem, whose function is to maintain pressure, comprises Active Vent Valves (AVVs), a Positive Pressure Relief Valve (PPRV) and a Negative Pressure Relief Valve (NPRV). The AVVs, which are essentially solenoid valves with manual overrides, are triggered when cabin pressure exceeds nominal levels. When this occurs, the AVVs open and simply vent to space.

Table 7.3 Crew Dragon Specifications

Length	2.9m
Diameter	3.6m
Sidewall Angles	15°
Pressurized Volume	10m³
Unpressurized Volume	14m³
Trunk Extension	34m³
Sensor Bay	0.1m³
Mass	4,200kg
Launch Payload	6,000kg
Return Payload	3,000kg
Endurance	Up to 2 Years
Maximum Crew	7
Avionics	Full Redundancy
Reaction Control	18 Draco Thrusters
Propellant	Hydrazine/Nitrogen Tetroxide
Propellant Mass	1,290kg
Docking Mechanism	LIDS or APAS
Power Supply	2 Solar Arrays – 1,500–2,000W
Power Buses	28V & 120V DC
Batteries	4 Li-Polymer Batteries
Cabin Pressure	13.9–14.9psi
Cabin Temperature	10–46°C
Cabin Humidity	25–75%
Command Uplink	300kbps
Downlink	>300Mbps
Windows	Up to 4
Window Diameter	30cm

The Oxygen and Nitrox System is a complex subsystem that has myriad functions. First, it stores oxygen that can be used in the event of a rapid decompression; second, it can provide breathing gas in the event of a contaminated

Figure 7.11. Crew Dragon interior, Credit NASA.

Figure 7.12. Crew Dragon's ECLSS. Credit NASA.

atmosphere; and third, the nitrox (a first in manned spaceflight) is used to cool the cabin air during reentry. The delivery of oxygen and nitrox to the cabin is via solenoid valves. One valve controls flow to the crew's suits and another valve controls flow to the cabin, which is pressurized to between 8.0 and 8.5 psia.

A third element of the ECLSS is the SpaceX spacesuit (this is not always considered a life support subsystem incidentally – it was part of the Shuttle's ECLSS but is not a part of the ISS ECLSS), which provides the wearer with protection from explosive decompression, inadvertent release of volatile organic compounds (VOCs) into the atmosphere, and a pressurized breathable atmosphere. Like all things SpaceX, the suit looks cool (Figure 7.11), which is an important requirement (ask anyone who works for SpaceX). The fully integrated suit features boots, gloves and a helmet that are permanently attached, unlike the Shuttle era's David Clark S1035 suit. The umbilical attached to the suit ensures a steady flow of air to cool the suit and delivers air and nitrox to the occupant.

The Air Revitalization System comprises a dehumidifier and various air distribution subsystems that help circulate the air while removing carbon dioxide and VOCs. Metabolic wastes are removed using four different types of air filters, which are in an air sanitation box. There is nothing cutting edge about the way carbon dioxide is removed from the cabin because Crew Dragon uses good old-fashioned lithium hydroxide canisters, which have been used since the Mercury Program in the 1960s (the ISS, by comparison, uses molecular sieves and advanced closed loop systems that recycle carbon dioxide and are much more efficient than lithium hydroxide, although the ISS stores a few lithium hydroxide canisters as back-up). Other contaminants such as flakes of skin and dust are dealt with using good old-fashioned off-the-shelf HEPA (High Efficiency Particulate Air) filters, like ones you can buy at your supermarket. Once air has been scrubbed of carbon dioxide and particulates, it is thermally conditioned using heat exchangers before being directed through the cabin, while airflow is achieved using cabin fans. Of course, with all these subsystems and assemblies working, a fair amount of heat is generated and this heat must be rejected. This is achieved using radiators, through which coolant is pumped.

Considering SpaceX had little to no experience in developing life support systems, the time between demonstrating the various subsystems and the ECLSS being certified operational was surprisingly quick, even for SpaceX. The first integrated demonstration of the ECLSS took place in November 2016. This four-hour test demonstrated ECLSS functionality under metabolic loading, thanks to test participants who provided metabolic waste products by exercising (this was done to increase carbon dioxide levels and humidity in the cabin). The flight test of the ECLSS took place during the Demo-1 flight in March 2019. Since this flight was unmanned, there was obviously no metabolic load (SpaceX could have flown a human patient mannequin that would have created a metabolic load, but decided against this for some reason), but the flight nevertheless returned noteworthy data that demonstrated system functionality, paving the way for human-in-the-loop testing that took place in January 2020. This test included four test participants who performed usual crew activities at the times they would be expected to do so during a mission. The activities performed by the test participants included swapping out lithium hydroxide canisters, responding to a simulated failure of the air distribution system and performing leak checks on the suits.

The ECLSS performed as advertised and the test set the stage for the first manned flight later that year. The Crew Dragon ECLSS not only served as a model for future SpaceX spacecraft, but its development served as a model – and a lesson – to spaceflight companies everywhere, of just how fast a system can be designed, developed and tested if you happen to have the right organizational structure in place. Thanks to a vertically integrated organizational structure, SpaceX teams can develop subsystems and assemblies at a rapid pace, because the personnel in those teams work so closely together. This results in a tight feedback loop, a byproduct of which is fast iteration through the testing and qualification process. It is a philosophy that many companies would do well to adopt. Incidentally, those interested in learning more about spaceflight life support systems are referred to the author's earlier work '*Life Support Systems for Humans in Space*', published in 2020 by Springer.

In addition to life support system testing, Crew Dragon was also subjected to flight tests, including a pad abort test and a hover test. The pad abort test, which took place at the SLC-40 launch pad leased by SpaceX, was completed on May 6, 2015, following a safe landing on the ocean 99 seconds after ignition. An instrumented crash test dummy wired to myriad sensors was the sole passenger on board during the pad abort, while the remaining six seats were loaded with ballast to approximate a full passenger load. Later that year, in November, SpaceX conducted a hover test in McGregor, Texas, to demonstrate the performance of the SuperDraco engines. Once the systems and subsystems of Crew Dragon had been developed and tested on the ground, all that remained was to conduct an in-flight abort test, which was performed on January 19, 2020. This flight test, which was conducted from LC-39A at the Kennedy Space Center (KSC) in Florida, sent the Crew Dragon on a suborbital trajectory. During the flight, separation and the abort scenario were evaluated at transonic velocity as the vehicle transited through Max Q. The SuperDraco abort engines were used to push the vehicle away from the Falcon 9 following a planned premature engine cutoff and, shortly after it detached, the Dragon followed a suborbital trajectory to apogee. The trunk was then detached, and the vehicle's Draco engines oriented the vehicle for descent. Following successful parachute deployment and landing, the data was reviewed and the flight was hailed as having met all flight test objectives.

The first test flight was Crew Demo-1, which flew to the ISS after launching on March 2, 2019. During the mission, the spacecraft tested approach and docking procedures with the ISS (the complete mission profile of a SpaceX manned mission is summarized in Tables 7.4 and 7.5).

Table 7.4: Crew Demo-1 Countdown Events

Hour / Min / Sec	Events
00:45:00	SpaceX launch Director verifies go for propellant load
00:42:00	Crew Access Arm retracts
00:37:00	Dragon launch escape system armed
00:35:00	RP-1 loading begins
00:16:00	2nd stage LOX loading begins
00:07:00	Falcon 9 begins engine chill prior to launch
00:05:00	Dragon transitions to internal power
00:01:00	Command flight computer to begin final prelaunch checks
00:01:00	Propellant tank pressurization to flight pressure begins
00:00:45	SpaceX Launch Director verifies go for launch
00:00:03	Engine controller commands engine ignition sequence to start
00:00:00	Falcon 9 lift-off

Table 7.5: Crew Demo-1 Launch, Landing and Deployment Events

Hour / Min / Sec	Events
+00:00:58	Max Q
+00:02:33	1st stage MECO
+00:02:36	1st and 2nd stages separate
+00:02:44	2nd stage engine burns
+07:07:15	1st stage entry burn
+00:08:47	2nd stage MECO
+00:08:52	1st stage entry burn
+00:09:22	1st stage entry landing
+00:12:00	Crew Dragon separates from 2nd stage
+00:12:46	Dragon nosecone open sequence begins

Life support systems were monitored during the flight, and G-loads were measured thanks to an Anthropomorphic Test Device (ATD – the technical specification is HYBIII-50M ATD) named Ripley (after the lead character in the classic 1979 science fiction film '*Alien*'). In addition to being fitted with accelerometers, Ripley (Figure 7.13) also sported ten sensors and various data acquisition systems to measure neck and spine data.

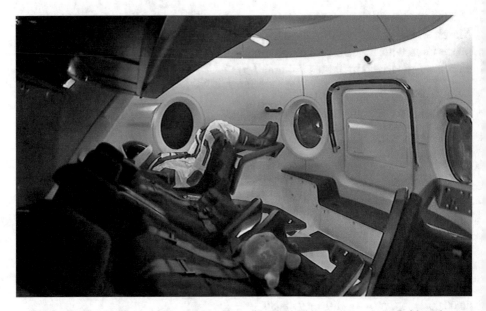

Figure 7.13. 'Ripley' relaxes inside Crew Dragon. This image was probably taken while on the ground. How do we know? That plush toy in the foreground is a super-high-tech zero-g indicator (it now resides on board the ISS) – a technique pioneered by the Russians many years ago. Credit NASA.

After successfully undocking from the ISS on March 8, 2019, Crew Dragon splashed down 320 kilometers off Florida's east coast the same day. The success of the mission set the stage for the first manned flight: Crew Dragon Demo-2.

Crew Dragon Demo-2 was important for several reasons, perhaps the most significant of which was the fact that the flight marked the first crewed orbital spaceflight to be launched from the United States since the final Shuttle flight in July 2011. Demo-2 also happened to be the first two-person flight launched since STS-4 in 1982. Unlike the Shuttle, which carried a 1-in-65 risk of failure, NASA calculated the Loss-Of-Crew (LOC) chance as a 1-in-276 risk (incidentally, the loss of *mission* risk – one in which the crew survives – was calculated as 1-in-60). As with all flights, Demo-2 had a shopping list of tasks to complete, including validation of crewed spaceflight operations to ensure the spacecraft could be approved by NASA for carrying astronauts. The docking and un-docking operations were conducted autonomously, but as with all manned spacecraft operations the crew had the option to intervene manually if necessary. The crew of Douglas Hurley and Robert Behnken (Figure 7.14), both Shuttle veterans, had been announced in August 2018, and their arrival at ISS, on May 30, 2020, increased the crew size from three to five (joining Chris Cassidy, Anatoly Ivanishin and Ivan Vagner).

Figure 7.14. Doug Hurley (L) and Bob Behnken, briefing mission controllers about their experience in Crew Dragon. The launch of Demo-2, which was attended by President Trump and Vice President Pence, was watched by 10 million people online, with 150,000 people – including the author – watching live from the Florida coast. Although the technical identifier of the vehicle was 'capsule 206', the name given to the spacecraft by the astronauts was *Endeavour*. One of the most important mission-critical items flown by the crew was an apatosaurus plush toy named *Tremor*, whose job it was to indicate to the astronauts when microgravity had been attained. Credit NASA.

Behnken and Hurley spent more than 100 hours conducting science experiments during their time on orbit, while Behnken also completed four EVAs together with Chris Cassidy. After spending close to 63 days on ISS, the vehicle undocked on August 1 and headed back to Earth after completing four departure burns and a six-minute phasing burn. During the re-entry phase of the mission, the crew was subjected to a maximum 4G. After splashing down safely, Crew Dragon was retrieved by the GO Navigator ship (Figures 7.15 and 7.16) and Behnken and Hurley were carted off on stretchers for medical assessment.

"Once we descended a little bit into the atmosphere, Dragon really came alive. It started to fire thrusters and keep us pointed in the appropriate direction. The atmosphere starts to make noise – you can hear that rumble outside the vehicle. And as the vehicle tries to control, you feel a little bit of that shimmy in your body... We could feel those small rolls and pitches and yaws – all those little motions were things we picked up on inside the vehicle...

All the separation events, from the trunk separation through the parachute firings, were very much like getting hit in the back of the chair with a baseball bat... pretty light for the trunk separation but with the parachutes it was a significant jolt."

Robert Behnken, describing the return from orbit.

Figures 7.15 and 7.16. Endeavor is hoisted on board the GO Navigator. Credit NASA.

Table 7.6: Inspiration4 Crew

Position	Astronaut
Spacecraft Commander (symbolizing Leadership)	Jared Isaacman
Pilot (symbolizing Prosperity)	Sian Proctor
Chief Medical Officer (symbolizing Hope)	Hayley Arceneaux
Mission Specialist (symbolizing Generosity)	Christopher Sembroski

We will conclude this chapter with another notable Crew Dragon flight. Inspiration4, the first all private citizen flight, was sponsored by Jared Isaacman, the founder and CEO of credit card processing company Shift4 Payments, who has a net worth of $2.6 billion. A pilot and graduate of Embry-Riddle Aeronautical University (the author's employer), Isaacman selected his crew based on four aims that aligned with the four attributes of leadership, hope, generosity, and prosperity (Table 7.6). Hayley Arceneaux, 29, (representing Hope) was a bone cancer survivor, and served as the crew's medical officer. The mission made her the youngest American in space. Sian Proctor, 51, (representing Prosperity) was a community college educator from Tempe, Arizona, who won her ticket through a competition held by the Shift4Shop eCommerce platform. The contest sought entrepreneurs worthy of being 'elevated to the stars'. The third seat went to Christopher Sembrowski, 41 (representing Generosity; the ticket was actually won by a friend who then gave his seat to Sembrowski).

> "Inspiration4's goal is to inspire humanity to support St. Jude here on Earth while also seeing new possibilities for human spaceflight. Each of these outstanding crew members embodies the best of humanity, and I am humbled to lead them on this historic and purposeful mission and the adventure of a lifetime."

Jared Isaacman, March 2021.

The crew of the SpaceX Crew Dragon *Resilience*, which flew from September 15 to 18, 2021, was the first to fly with no professional astronauts on board. SpaceX spent six months training the crew, which included a climb on the slopes of Mount Rainier and a parabolic flight on board a Zero Gravity Corporation aircraft. The most qualified of the crew was Isaacman, a jet pilot and founder of Draken International, a defense company that trains pilots. The *Resilience* (a repeat flyer), which lifted off at 8.02 pm, EDT on September 15, was placed into a 576-kilometer orbit. Thanks to a cupola, the crew had plenty to time to soak up the view in between participating in air-to-ground broadcasts and ringing the closing bell of the New York Stock Exchange (on September 17). Incidentally, for those of you like records, the arrival of the Inspiration4 crew on orbit set a new record of 14 astronauts being on orbit simultaneously. The crew returned to Earth on September 18, with the *Resilience* making an autonomous descent that started by firing the spacecraft's Draco engines for 15 minutes beginning at 6:20 pm EDT. The *Resilience* splashed down at 7:06 pm EDT, SpaceX recovery teams recovered the capsule 30 minutes later, and the crew flew back to the Kennedy Space Center via helicopter.

8

Red Risks

"Ultimately, the thing that is super important in the grand scale of history is 'Are we on the path to becoming a multi-planet species or not?' And if we're not, that's not a very bright future; we'll simply be hanging out on Earth until some eventual calamity claims us."

Elon Musk, speaking at a conference held on August 2, 2011, by the American Institute of Aeronautics and Astronautics.

As *Curiosity*, the $2.5 billion Mini Cooper-sized Mars rover, touched down on the Red Planet on August 3, 2012, Elon Musk was already planning the next logical step – sending humans there. As with all of Musk's space plans, his goal was not short on ambition. The Mars Messiah is not interested in merely ferrying people to Mars; Musk wants to make it possible for people to live there. Permanently. Musk acknowledges that one of the biggest challenges of colonizing the Red Planet is making the trip affordable, suggesting a round-trip ticket price should be around half a million dollars. It is a bold plan, but 'bold' is an appropriate moniker for the man who created PayPal, Tesla, and SpaceX.

Manned missions to Mars have always been a popular spaceflight theme, and not just in the wake of the arrival of *Curiosity*, and more recently, *Perseverance*. In September 2011, NASA unveiled the Space Launch System (SLS) that the agency hopes will deliver humans to the Red Planet by the early 2040s. Then there was that controversial Mars One project, which planned to send humans on one-way trips to Mars starting in the 2020s. The Mars One project, which is now bankrupt, planned to use SpaceX hardware to transport their astronauts to the Red Planet. But despite all the discussion about sending humans to Mars, there have been few more vocal than Musk, who believes his company can land humans on the Red Planet as soon as 2028. If he is right, it is entirely possible that *Curiosity* and *Perseverance* could still be scuttling around when the first Musketeers land.

© Springer Nature Switzerland AG 2022
E. Seedhouse, *SpaceX*, Springer Praxis Books,
https://doi.org/10.1007/978-3-030-99181-4_8

Landing humans on Mars by the end of the 2020s is as ambitious as it sounds. In the weeks leading up to the arrival of *Curiosity* and *Perseverance*, the public was inundated with press releases from NASA explaining just how difficult it is just to land a robot there. Getting a human there is a completely different ballgame. For one thing, humans require bulky life support systems, food and living space. Then there are the myriad physiological hazards of long-duration space travel. Musk understands the inherent dangers but, while he acknowledges the flights will be risky, he also has no qualms about being on the first flight so long as he feels comfortable he has left SpaceX in the right hands… just in case.

> *"It's dangerous, it's uncomfortable, it's a long journey. You might not come back alive. But it's a glorious adventure, and it'll be an amazing experience. You might die… and you probably won't have good food and all these things. It's an arduous and dangerous journey where you may not come back alive, but it's a glorious adventure. Sounds appealing. Mars is the place. That's the ad, that's the ad for Mars."*

Elon Musk, being interviewed by Peter Diamandis, April 26, 2021.

If it was anyone else talking about how to send people to Mars for half a million dollars each then one would have to be skeptical, but when the person talking is Elon Musk it is difficult to be anything but guardedly optimistic, especially when you consider the track record SpaceX has for getting stuff done. After all, Musk's company got the vote of confidence from NASA, which entrusted SpaceX with returning American astronauts to space on the space taxi mission launched in 2020. The question is, how soon can SpaceX follow ISS ferry flights with a Mars crewed mission? One way of achieving Musk's interplanetary dream lies in the SpaceX principle of eliminating some of the equipment costs for space travel. You have no doubt seen images of Falcon 9 stages landing autonomously; by increasing this reusability and leaving fuel as the only financial burden, SpaceX is on track to make spaceflight more affordable, and the reusability principle is one that can be applied to a manned Mars mission.

In the early days of manned Mars mission planning, SpaceX intended to use a Falcon Heavy to launch an unmanned sample return mission to test the techniques and technologies required for a manned mission. The sample return mission would have used a Dragon variant – the appropriately named Red Dragon – to ferry instruments such as a drill to penetrate underground to sample reservoirs of water ice (potential landing sites would have been polar or mid-latitude sites, with proven ice fields known to exist below the surface). NASA's Ames Research Center (ARC) got involved with SpaceX to plan a mission that was projected to cost less than US$400 million, plus $130 million for the Falcon Heavy. In addition to the search for life, assessing subsurface habitability and establishing the distribution of ground ice, the Red Dragon mission would have conducted a human-relevant Entry, Descent and Landing (EDL) test, demonstrated access to subsurface resources, and conducted In-Situ Resource Utilization (ISRU) tests, all important technologies required in a manned mission. Ultimately, work on Red Dragon was discontinued in July 2017 and resources directed to development of the Starship. But before we delve into the details of Starship in the next chapter, just how risky will this much-hyped Mars mission be?

One of the risks we often hear about whenever a rover is about to land on Mars is the 'Six Minutes of Terror', a reference to the EDL problem. While this is a significant risk, perhaps a more challenging threat is the effect of long duration spaceflight on astronaut health. In short, the myriad physiological hazards of a deep space mission must be controlled and mitigated to ensure that it is successful. These hazards (Figure 8.1), which include bone loss, muscle atrophy, radiation exposure, isolation and several others, are generally ranked using a color-coded scheme to indicate the severity of the risk (Figure 8.2). For example, risks ranked as red, such as radiation exposure and bone loss, have the most serious impact on human health and performance. There are some who argue that because astronauts routinely spend six months on the International Space Station (ISS) we have already proven that astronauts can travel to Mars safely, because six months is about the length of time it takes to travel to the Red Planet. The problem is that once astronauts arrive at Mars they must then work, live and survive there for several months before making another six-month trip back. That is a problem because the astronauts are practically invalids on their return to Earth after only six months on the ISS (Figure 8.3), having lost seven percent of their bone mass and 20 to 25 percent of their muscle mass, resulting in a 30 percent loss of force. Now imagine what those numbers will be after more than two years in space.

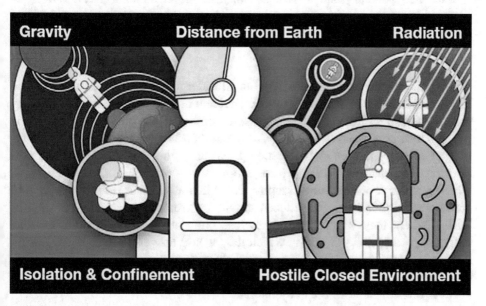

Figure 8.1. Primary threats to physiology associated with long duration spaceflight include exposure to radiation, reduced gravity, hostile environments, distance from Earth, and isolation. Courtesy NASA.

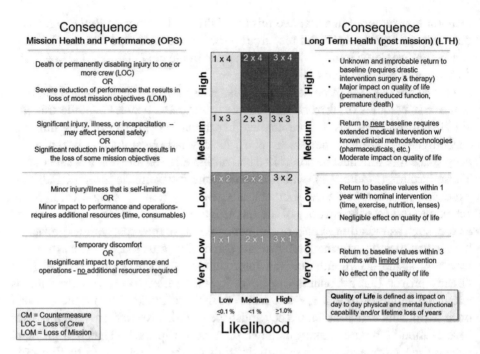

Figure 8.2. NASA's Human Research program (HRP) applies a research-based approach to calculate mission risks to astronauts, as depicted in this matrix. On the left are the consequences of the risks during the mission and on the right are long-term health risks. Risks shown in red are rated for their likelihood of causing long-term health issues, while risks shown in yellow are medium level risks (these risks will require monitoring during the mission). Risks shown in green are those that can be controlled through application of mitigation strategies. Courtesy NASA.

We will get into the details shortly, but consider just one of these hazards: bone loss. Astronauts lose between 1.0 and 1.2 percent of their bone mass density every month on orbit with no clinical horizon (ten times the rate at which osteoporosis patients lose bone mass incidentally, or about the same amount of bone the average person loses in their sixth decade). That means that at the end of six months on orbit the astronauts will have lost about seven percent of their bone mass, which equates to a three-fold fracture risk. Mars astronauts would then have to spend several months in the reduced gravity environment of Mars (0.38 that of Earth), where they would continue to lose bone mass, before making the six-month trip back to Earth during which they would lose even more bone. Losing bone is just one of the many physiological stressors astronauts must contend with. Compounding the problem is the fact that these stressors do not act independently, because their impacts on the body interact.

Figure 8.3. After nearly six months on the ISS, Expedition 36 astronaut Chris Cassidy (L) and cosmonauts Pavel Vinogradov and Alexander Misurkin (R), recover in couches outside the Soyuz TMA-08M capsule after landing in Kazakhstan, September 11, 2013. Courtesy NASA.

Before continuing with the subject of bone loss, we should mention the hazard presented by radiation. Of all the challenges SpaceX astronauts will face, radiation is the most damaging. Part of the reason is that the body cannot adapt to radiation and the longer an astronaut is in space, the more radiation they will be exposed to. As we shall see, too much radiation can have significant negative effects on human physiology. But first, a primer. There are two types of space radiation (Figure 8.4): galactic cosmic rays (GCRs) and solar particle events (SPEs). GCRs are unique because they are nigh on impossible to shield against. They are highly energetic charged particles that originate outside the solar system and bullet along at close to the speed of light, propelled through space by the force of exploding stars. This means these particles have tremendous energy. The mass of some of these particles, such as iron, combined with their phenomenal speed, enables them to slice through the walls of a spacecraft like the proverbial hot knife through butter. Astronauts exposed to too much cosmic radiation are at higher risk of developing cancer. Here is what Dr. Francis Cucinotta, one of the world's leading experts on the physiological effects of space radiation, has to say:

"The type of tumors that cosmic ray ions make are more aggressive than what we get from other radiation."

Dr. Francis Cucinotta, University of Nevada, Las Vegas

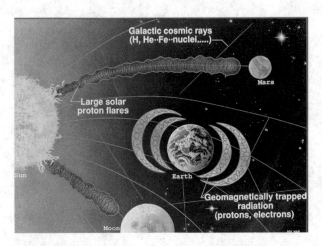

Figure 8.4. Space radiation. Credit NASA.

But how much is too much? Before we answer that we need to know how radiation is measured, and the usual metric is the Sievert. Here on Earth your annual radiation dose is about 2 milliSieverts (2 mSv), whereas astronauts spending six months on board the ISS are exposed to about 80 mSv. A comparative metric is the chest radiograph dose, which is 0.02µSv per hour. If you happen to spend a lot of time flying commercially, the radiation dose per hour is about 0.3 to 5.7µSv. Spend too much time on orbit and astronauts may hit their career radiation limit, which equates to an increase of three percent risk of developing cancer in their lifetime (see sidebar: *Radiation Risk*). Any astronaut heading for Mars will be exposed to 1.84 mSv per day during the outbound and inbound transit phases and 0.7 mSv per day on the surface, which equates to a mission total of about 1.5 to 2.0 Sv.

Radiation Risk

In the United States, the incidence rate of cancer is 38.5 percent (according to the National Cancer Institute – www.cancer.gov – based on statistics between 2008–2012). If you exposed 100 people (which is the capacity of the SpaceX Starship, incidentally) to the amount of radiation that Mars astronauts will be exposed to, 61 of them would be diagnosed with cancer. By virtue of the unique characteristics of GCR, these cancers would typically be lung, breast and colorectal cancers, meaning half these astronauts would die. Scientists have modeled the dangers of GCRs during a manned Mars mission and have calculated that exposure to radiation on such a trip would shorten an astronaut's lifespan by between 15 and 24 years.

The biological effects of exposure to space radiation can be divided into *acute* and *chronic*. Acute effects are the result of exposure to high radiation doses, which may be caused by SPEs, whereas chronic effects (Figure 8.5) are caused by extended exposure to space radiation. The potential effects of either type of exposure include *direct* and *indirect* damage to genetic material, biochemical alterations of cells and/or tissues, carcinogenesis, degenerative tissue effects and cataracts. The extent of these effects is determined by the type of radiation, its flux and the energy spectrum, factors that are not completely understood. Other factors that determine radiation damage include age at exposure, gender and susceptibility to radiation. The quantitative physiological effects of radiation are also poorly understood, due partly to misinterpretation of the exact mechanisms and processes that concern DNA repair. Another poorly understood mechanism is that of Relative Biological Effectiveness (RBE), which is determined to a large degree by radiation type and kinetic energy. How RBE correlates with tumor type or cancer progression is practically unknown, because most of the limited experimental data has been conducted on mice and it is difficult to extrapolate and apply mice data to humans. It is even more of a challenge to use that data to estimate health risks for cancer, cataracts, and Central Nervous System (CNS) risks. To gain even a cursory insight into the problem will require considerably more astronauts conducting one-year (or longer) increments followed by lengthy post-mission observation times. Given that the ISS is due to be retired in 2028, this goal is currently impossible. Even if it was possible, extrapolating data from the ISS to deep space is extremely limited at best in terms of making accurate risk predictions for those on the Starship who will be venturing beyond Earth orbit. In short, there are myriad knowledge gaps regarding the potential acute and late biomedical risks from GCRs and SPEs, but what follows is some of what we do know.

"I'm having these light flashes. I'm seeing this, like, light flashing in my eyeballs.
It was like fireworks in your eyeballs. It was spectacular."

Charles Duke, Lunar Module Pilot, Apollo 16

We will begin with CNS risks. The potential acute and late risks to the CNS from GCRs and SPEs have not been a major consideration for ISS crews because these astronauts are exposed to low doses of ionizing radiation compared to deep space doses. The risk presented by GCR was evident during the Apollo era, when astronauts reported the 'light flash' phenomenon caused by cosmic radiation traversing through the retina. This phenomenon is routinely reported by ISS astronauts, incidentally. As these particles of cosmic radiation traverse, they cause microlesions. In addition to microlesions in the eye, exposure to GCR, at the kind of dosage astronauts will be exposed to during a Mars mission, causes immune system compromise and wearing away of the myelin sheath that protects nerve

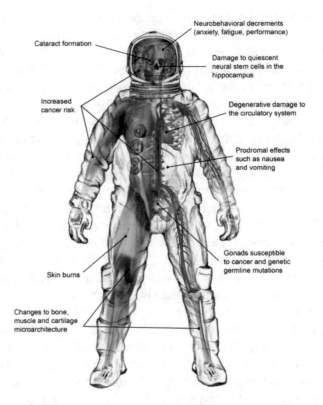

Cataract formation

Neurobehavioral decrements
(anxiety, fatigue, performance)

Damage to quiescent
neural stem cells in the
hippocampus

Increased
cancer risk

Degenerative damage to
the circulatory system

Prodromal effects
such as nausea
and vomiting

Gonads susceptible
to cancer and genetic
germline mutations

Skin burns

Changes to bone,
muscle and cartilage
microarchitecture

Figure 8.5. Some of the risks SpaceX astronauts will face from radiation. Credit NASA.

fibers. Other CNS risks include detriments in short-term memory and altered motor function. For example, one radiation-related topic that has received media attention in recent years is the suggestion that exposure to deep space radiation may cause cognitive deficits. One study that investigated this exposed rats to 1,000 MeV/u before testing their spatial memory in a radial maze. In this study, the exposed rats committed more errors than control rats and were unable to develop a spatial strategy to make their way through the maze. A similar study examined mice that had been subjected to two weeks of whole-body irradiation. The results of this study revealed impaired novel object recognition and reduced spatial memory (Figure 8.6). Another CNS effect is altered neurogenesis.

Neurogenesis is a term that describes the formation of neurons. In adults, this process occurs in the Sub-Ventricular Zone (SVZ) and the sub-granular zone of the brain. Scientists are still researching the role that neurogenesis plays in cognition. Since the formation of neurons may be sensitive to radiation, it is possible that long-term exposure may result in cognitive deficits.

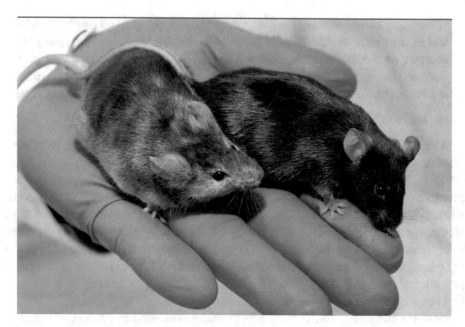

Figure 8.6. Studies using mice have shown that radiation exposure during a Mars mission may cause cognitive deficits. Credit NASA.

On the subject of cognitive deficit, it is important to mention oxidative[1] stress as this is thought to be implicated in Alzheimer's disease. Because radiation has been shown to increase oxidative damage, oxidative stress represents yet another mechanism of a radiation-induced cognitive injury. Since antioxidants prevent such damage under normal conditions, it would seem logical to suggest that astronauts eat food containing elevated levels of antioxidants. For example, a diet high in blueberries or strawberries should help offset oxidative stress. Melatonin is another option. Melatonin has high antioxidant properties and studies have shown it inhibits neurogenesis. The problem with research to date is that studies have used high dose rates and the biological effects of radiation are different at low dose-rates. Furthermore, studies that have investigated the beneficial effects of antioxidants found no evidence that supplementation is effective. In fact, some studies revealed that antioxidants such as vitamin A, vitamin E and β-carotene might be more damaging, because taking extra amounts of antioxidants would help the body rescue cells that had been damaged by radiation and this might alter DNA repair.

---

[1] Oxidative stress is a term that describes the imbalance between the production of free radicals and the ability of the body to neutralize these free radicals using antioxidants. Free radicals are molecules that contain oxygen. These molecules have one or more unpaired electrons, which means they are very reactive with other molecules, which in turn means they are capable of chemically interacting with and destabilizing cells such as DNA. Under normal conditions, antioxidants prevent these reactions.

"This study shows for the first time that exposure to radiation levels equivalent to a mission to Mars could produce cognitive problems and speed up changes in the brain that are associated with Alzheimer's disease. These findings clearly suggest that exposure to radiation in space has the potential to accelerate the development of Alzheimer's disease."

Dr. Kerry O'Banion, University of Rochester Medical Center.

Another CNS-linked risk is Alzheimer's, which is a neurodegenerative disease that causes dementia in most cases. Common symptoms include short-term memory loss, language problems, disorientation, lack of motivation, and behavioral problems. Alzheimer's, which is chronic, begins slowly and symptoms become worse with time. The cause of the disease is not completely understood, but a large element of the risk is believed to be genetic. It can also be precipitated by exposure to radiation and it has been shown in mice that exposure to radiation accelerates the onset of age-related neuronal dysfunction that results in symptoms like those exhibited by people suffering from Alzheimer's. There are currently no treatments that can stop the disease or even slow its progression.

Now we can return to the bone loss problem. Yet another process affected by radiation is bone remodeling. For astronauts in Low Earth Orbit (LEO), the rate of Bone Mineral Density (BMD) loss is predictable, but beyond LEO the radiation environment is harsher, and the rate of BMD loss is less predictable. This is because bone is damaged more by higher doses of radiation, a fact long documented by the persistent decline in bone volume following exposure to therapeutic radiation in cancer patients.

Compounding the effect of osteoradionecrosis (the term used to describe radiation-induced bone loss) is the effect that radiation has on fracture sites. We have a good understanding of this process because ionizing radiation has long been used as a treatment for malignancies and has been a factor in reducing cancer mortality. One of the main reasons BMD is reduced following irradiation is because osteoblasts and osteoclasts are damaged (osteoblasts and osteoclasts are two bone cells that remodel bone; osteoclasts break down bone and osteoblasts build up bone). When these bone cells are damaged, bone formation is impaired due to cell-cycle arrest. One of the processes by which osteoclasts and osteoblasts are damaged is oxidative stress caused by radiation, since it is this oxidative stress that damages osteoprogenitors. To begin with, irradiation causes an increase in osteoclast number which causes osteoporosis. Shortly after exposure there is a decline in the number of osteoclasts and osteoblasts, which results in suppression of bone remodeling and degradation of bone quality. But the side effects of this treatment have concerned oncologists and will be of concern to flight surgeons responsible for the health of astronauts embarked on Mars missions. This is because bones within the irradiated area are at a much higher fracture risk.

For example, patients undergoing breast cancer treatment may have rib fracture rates that exceed 15 percent. This will be of concern to Mars-bound astronauts because their bones will already be weakened due to the loss of BMD simply from being in microgravity. Consequently, these astronauts may be at elevated risk of traumatic and/or spontaneous fracture.

So, the effect of radiation results in a cascade of changes. In addition to the reduction in BMD and the impact on healing, the way in which bone is weakened will be of concern to flight surgeons tasked with keeping a crew fracture-free. For example, the loss of trabecular bone means cortical bone must now deal with a greater proportion of loads on the skeleton. This in turn means that cortical bone will be increasingly less able to resist the torsional and bending loads, a change that may be exacerbated by any defect in the bone such as a porous hole. The net effect of all these radiation-induced effects is an overall disruption of load distribution that results in compromised structural integrity of the bone. For astronauts about to land on Mars this is not an optimum situation. It is important to remember that astronauts return to Earth with increased fracture risk at the end of their ISS increments, but this does not mean a fracture is imminent. It just means that there is a greater chance of fracture after their return due to the loss of BMD. But this increased fracture risk is dramatically reduced thanks to the rehabilitation schedule that astronauts follow on return to Earth. This results in regeneration of bone, a process that typically takes between three to five years. In addition to osteoradionecrosis there is the condition of osteitis, in which the bone's ability to withstand trauma is reduced. In this condition, non-healing bone may be susceptible to infection, and the ability of the bone to heal is further complicated by hypovascularization. As the body is subjected to increased radiation, the small blood vessels inside the bone are destroyed. This is devastating because it is these blood vessels that carry nutrients and oxygen to the bone. Without blood vessels to do this, the bone simply dies.

So, we have covered the issue of bone loss and some of the radiation risks related to bone. Now we need to discuss the challenges of maintaining bone health, but first a review for any non-physiologists. Bone plays several important roles. First, it serves as a structure that supports the body. Second, it stores calcium. Third, it produces blood. Bone is also very dynamic. This is because of the process known as bone remodeling (Figures 8.7a and 8.7b). This process relies on the activity of those two particularly important bone cells; osteoblasts that build up bone and osteoclasts that break down bone. On Earth, your skeleton undergoes a tremendous amount of loading. Those of you who wear Fit Bits will know you take between 8,000 and 10,000 steps every day just performing daily activities. That loading is detected by your brain and signals those osteoblasts and osteoclasts to go to work.

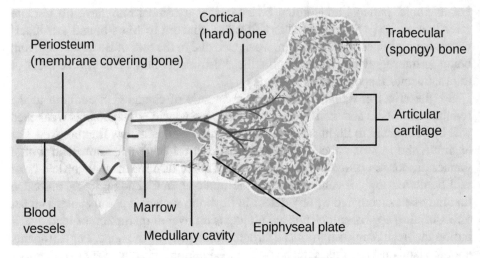

Figure 8.7a.   Bone structure. Open source.

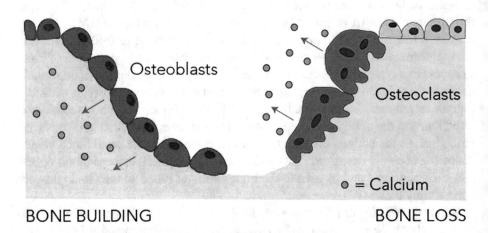

Figure 8.7b.   Osteoblasts and osteoclasts. Open source.

The result is a healthy skeleton that is very fracture resistant. In space, on the other hand, almost all of that loading is removed and the astronaut's brain detects that reduction in load and adapts accordingly. Less load means that strong bones are no longer needed, so the process of bone remodeling is readjusted. In short, the action of the osteoblasts is decreased and the action of the osteoclasts is increased, resulting in that reduction in bone mass density of between 1.0 and 1.2 percent every month. That is ten times the rate at which osteoporosis patients lose bone. Yes, we have mentioned this before, but it is important to emphasize just how serious a problem bone loss is (Figure 8.8). The body not only alters the responses of the osteoblasts and osteoclasts, it also alters calcium balance. Calcium is a key bone building material, but in microgravity it is not necessary to have strong bones, so the body decides to get rid of calcium to the tune of ~250mg per day. Not only does calcium excretion result in weaker bones, it also increases the risk of kidney stones because that calcium must be routed through the kidneys before being excreted. But worse is to come. Take another look at Figure 8.3. This crew has just returned from an ISS increment lasting several months. Notice anything particular in this image (apart from the fact they are on their cellphones)? They are *horizontal*. Why? Because their bones are so weak after having spent so long in space. Some of the people milling around them in the image are flight surgeons entrusted with the care of the crew. Shortly after landing, the crew are whisked away to their respective agencies and will embark on a dedicated and personalized rehabilitation program to regain lost muscle and bone. After about 90 days, most astronauts have recovered their muscle, but the bone loss is another story. There have been some astronauts (ESA astronaut, Thomas Reiter, following his 1995 Mir mission, for example) who still had not completely recovered their bone loss ten years after their six-month increment in space. You see, some astronauts recover their bone mass (relatively) quickly (three to four years) and other astronauts take longer.

Another aspect of this bone loss is bone mass density and architecture. The rod-shaped structures you can see in Figure 8.8 help the *architecture* of the bone. The better the arrangement of those rods and plates, the better your architecture and – theoretically – the stronger your bones. But another element that has a bearing on bone strength is your BMD. You can have fantastic bone architecture but if your bone density is low then your bones will be weak, and vice versa. The ideal scenario is to have both good bone density *and* good bone architecture (choose your parents carefully if you have ambitions to be a Mars-bound astronaut because part of the selection procedure may require enhanced medical screening). In space, the body remodels bone in response to the load demands on the bone, and since those load demands are low because of the absence of gravity, bone density is reduced. Those rods and plates are arranged to deal with the reduced loads in microgravity, which means weaker bone architecture. In LEO, the effect is manageable thanks to limited time on orbit and the personalized rehabilitation programs waiting for astronauts when they return to Earth. But a Mars mission will require astronauts to spend the best part of 12 months in deep space plus a surface stay in the Red Planet's reduced gravity environment. Imagine a crewmember suffering a femoral fracture. How would the crew cope? Could they cope? Astronauts are a resourceful lot, and they could probably cope for a while, but sooner rather than later the demands of caring for a disabled astronaut would exceed the limits of crew health care resources. Look at Figure 8.9. The image depicted is one way of dealing with a long bone fracture. The technique is known as *external fixation*. Imagine all the challenges of dealing with this in a reduced gravity environment: bleeding,

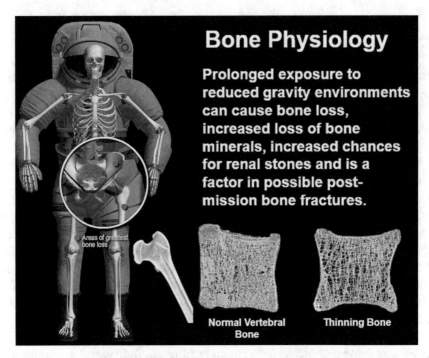

Figure 8.8. Bone physiology. Credit NASA.

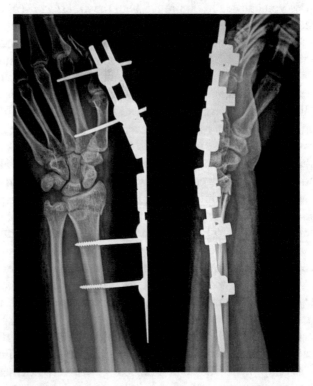

Figure 8.9. External fixation. Imagine a crew having to deal with this situation on the surface of Mars. Open source.

infection, re-infection, sepsis, 24/7 care, etc. So, what can we do? Well, there is a strategy called 'countermeasures'. It is a strategy that can be applied to limiting loss of BMD *and* reducing the rate at which muscles atrophy, so before we discuss what these countermeasures are we need to understand what happens to muscles during time in space.

There are three types of muscle in the body. Smooth muscle, also categorized as involuntary muscle, is found in organs and organ structures. Cardiac muscle, which is also involuntary muscle, is found only in the heart. Skeletal muscle, also categorized as voluntary muscle, is used for movement. You have about 640 skeletal muscles in your body. Skeletal muscle (Figure 8.10), also known as striated muscle due to its appearance under the microscope, helps support the body, assists in bone movement, and protects internal organs. It is divided into various subtypes. Type I, also known as *slow twitch* (red), is dense with capillaries, which means these muscles can carry a lot of oxygen and sustain lengthy periods of aerobic activity. Type II, also known as *fast twitch*, can sustain short bursts of activity.

Now we should examine some basic exercise physiology. Despite what you may hear in the gym, muscle cannot be turned into fat and fat cannot be turned into muscle. Second, the number of muscle fibers cannot be increased, no matter

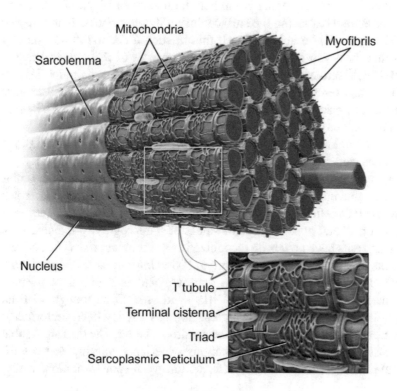

Figure 8.10. Skeletal muscle structure. Open source.

how hard you exercise, so how do muscles get bigger? Simple: the muscle cells get bigger in a process known as *hypertrophy*. This is the opposite of *atrophy*, which is the term that describes the wasting away of muscles. In skeletal muscle, movement is achieved by contraction stimulated by nerve impulses at a site known as the neuromuscular junction. Energy for muscle movement is in the form of glycogen, which is stored in the muscles and in the liver. When exercising, the muscles contract through actin and myosin filaments sliding over one another, a mechanism known as the sliding filament theory.

On Earth, muscle tone is maintained by exercise. Even sedentary people will take 8,000 to 10,000 steps every day, which is usually sufficient exercise to maintain some muscle tone. But in space, muscles do not get used as much as on Earth. The result is that muscles begin to atrophy at a rapid rate. In just six months, astronauts will lose about 25 percent of their total muscle mass, but this loss is not evenly distributed. A larger proportion of muscle loss is in the load-bearing muscles (your big leg muscles) and the balance muscles (those muscle groups that support your spine for example). As astronauts spend more time in space, their muscle cells shrink and become smaller and smaller. As those muscles become smaller they also become weaker and weaker, which means astronauts cannot exert as much force. Why is this important? Let us think operationally, referring to all those images and videos we have been bombarded with of astronauts going about their business bouncing around on the surface of Mars; of astronauts building outposts and exploring the surface during lengthy EVAs. Is it really conceivable that they will be able to do all that in such a deconditioned state? Do not forget that these astronauts will have lost a quarter of their muscle mass, which includes their cardiac muscle mass. This means the heart is much less efficient, which means exercise capacity is reduced. As mission time increases, astronauts will become weaker and weaker and... well you get the picture. To help mitigate this situation, countermeasures (Figure 8.11) can be applied.

By now, we know that astronauts' bodies suffer in microgravity. But it is not only bones and muscles that are affected. Eyesight may also suffer. We have known about vision impairment in astronauts for some time, but the problem has only been put under the spotlight recently after some astronauts experienced severe eyesight deficiencies. Thanks to anecdotal reports by astronauts and a comparison of pre- and post-flight ocular measures, microgravity-induced visual acuity impairments have now been recognized as a significant risk (you can read more about this in Springer's *Microgravity and Vision Impairments in Astronauts* by this author). The problem has its own acronym and is referred to as the Visual Impairment/Intracranial Pressure (VIIP) syndrome. Even though VIIP has only recently been identified, significant research has already been performed, so scientists are beginning to understand the syndrome better. The data shows that astronauts who suffer VIIP-related symptoms experience varying degrees of visual performance decrements. Some suffer cotton-wool spot formation, while others

Figure 8.11. Exercise countermeasures. Top Row (L to R): The advanced resistive exercise device (ARED). The T2 treadmill. The Cycle Ergometer with Vibration Isolation and Stabilization System (CEVIS). Middle Row (L to R): Interim Resistive Exercise Device (iRED) on ISS. The cycle ergometer aboard the Shuttle. Treadmill with Vibration Isolation and Stabilization System (TVIS) in the Zvezda Module on ISS. Bottom Row (L to R): The 'Apollo Exerciser'. The Teflon-covered treadmill-like device used during Skylab 4. Credit NASA.

may present with edema of the optic disc (Figure 8.12). Other astronauts may suffer flattening of the posterior globe, while some may present with distension of the optic nerve sheath. In short, there is a profusion of signs and symptoms but the reason for the vision impairment still has researchers a little flummoxed.

One theory suggests that the changes in ocular structure and impairment to the optic nerve are caused by the cephalothoracic fluid shift that astronauts experience while on board the ISS. It is theorized that some astronauts are more sensitive to fluid shift due to genetic and anatomical factors. You see, when you arrive on orbit there is a shift of fluids towards the head and a redistribution of body fluids. The body has about five liters of blood in addition to other body fluids, such as interstitial fluid which is found between the organs, and CSF which flows in the spinal cord. When astronauts arrive on orbit, between 1.5 and 2.0 liters of this fluid

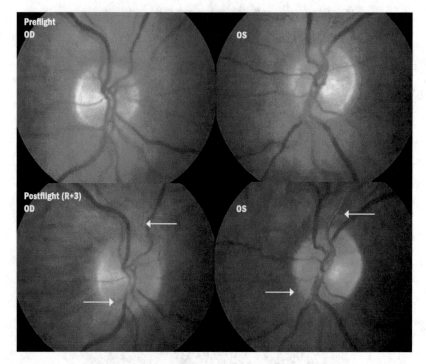

Figure 8.12. Cotton wool spots. Credit NASA

moves from the lower extremities to the chest and head. Not surprisingly, this causes various signs and symptoms, including facial puffiness, 'bird-leg' syndrome, pounding headaches and the vision problems mentioned earlier.

This fluid shift also triggers a series of adaptive processes in the body (remember, the body will always try to adapt to the environment). Inside your body, you have all sorts of sensors and receptors that send the brain information about temperature, electrolyte balance, and pressure. The pressure receptors are termed *baroreceptors* and, if fluid is maintained within certain thresholds, no action is taken. But when up to two liters of fluid is translocated from the lower to the upper body, this causes a spike in pressures that exceed thresholds. So, baroreceptors send this information to the brain and the brain decides that something must be done and begins the process of reducing pressure. The way it does this is to get rid of the excess fluid by triggering suppression of the renin-angiotensin-aldosterone system. This releases atrial natriuretic peptide which results in *diuresis*, a physiological term meaning you have to visit the washroom frequently to urinate. Unfortunately, all this urination has the side-effect of reducing plasma volume. About 55 percent of your blood is plasma and about 90 percent of your plasma is water, so if you are urinating frequently you are losing body water and hence, blood volume. In fact, in the first 24 hours on orbit, astronauts lose 17 percent of their plasma volume, which equates to an overall reduction of about 10 percent of total blood volume. The body does its best to adapt, which it does after about six weeks, but this adaptation is to microgravity, not 1G. On returning to Earth (or Mars), all that fluid that was in the upper body rushes to the lower body, which causes orthostatic intolerance (inability to stand upright); 25 percent of astronauts returning from space cannot stand upright for ten minutes within hours of landing because of orthostatic intolerance. This is one of the reasons that NASA prohibits astronauts from driving for three weeks postflight.

"Men Wanted for Hazardous Journey. Small wages, bitter cold, long months of complete darkness, constant danger, safe return doubtful. Honor and recognition in case of success."

Advertisement for Ernest Shackleton's 1914 Imperial Trans-Antarctic Expedition.

All we have discussed so far provides an insight into the physiological challenges SpaceX astronauts will face, but what about the psychosocial challenges? This is the least challenging aspect of Musk's Mars mission, because astronauts possess a wide-ranging repertoire of behavioral competencies that help them function effectively in a multi-cultural environment. Much like the bold explorers who

Figure 8.13. The crew of the *Endurance*, perhaps the greatest exploration vessel in history, strike out to escape their icy prison in the Southern Ocean. With their ship crushed and sunk by the Antarctic ice, Shackleton's crew was forced trek across towards the far end of the Antarctic Peninsula, taking with them only what they could drag along in their three lifeboats. Credit National Geographic.

answered the call to be a part of Ernest Shackleton's expedition (Figure 8.13), this repertoire is critical because a spacecraft is an environment in which faults cannot be tolerated. It should not come as much of a surprise that Musk reckons the pitch for the first astronauts to venture to Mars will echo the requirements of Shackleton's expedition. In other words, the journey will be diabolically difficult, extremely dangerous, and there will be a good chance many will die, but those that survive will reap the rewards.

Nowadays, to ensure astronauts have the skill sets required to cope with long duration expeditions, space agencies apply specific 'select-out' and 'select-in' criteria during selection. Once selected, astronauts have to complete extensive pre-flight training to develop 'expeditionary behavior'. This training comprises myriad space-related psychosocial skills designed to ensure mission success. In addition to all this preparation there is a ground-based complement of support staff who provide behavioral support via video conferences, to ensure missions proceed smoothly. We will assume SpaceX will be applying similar criteria to training its cadre of astronauts, but the process of selecting astronauts remains a challenging task. Interpersonal dynamics and difficulties, crew performance breakdown, and human interaction and performance in a confined and isolated environment are just a few of the factors that must be considered. Additional selection criteria such as communication competence, along with intercultural training, also have a decisive impact on future mission success.

It is for these reasons that the psychological aspect is often identified by the media as one of the more problematic issues affecting long duration crews. Nothing could be further from the truth. You see, there is a wealth of information about selecting crews for arduous expeditions. Take Shackleton. Nearly 5,000 men applied for the 27 jobs available on the great explorer's Imperial Trans-Antarctic Expedition (by comparison, NASA received just 3,654 applications for its 2009 astronaut selection campaign – and this was in the age of the internet remember). Shackleton personally interviewed each candidate whom he felt had potential. While he obviously had to have crewmembers who had sailing and scientific skills, he also wanted people who had the enthusiasm and optimism to help cope with expedition demands. Fortunately, Shackleton had an eye for talent and knew how to build a team that could survive just about anything. Each crewmember was selected to do a specific job and each man did his job well, which is probably why Shackleton's team survived for more than two years in Antarctica when all seemed lost.

In common with a manned Mars mission, Shackleton's expedition was a dangerous one (arguably much more dangerous than a Mars mission) and the success of his mission depended on a good team. His ideal crewmember had to be qualified for work on board the *Endurance*, but they also had to have special qualifications to deal with the extreme (polar) conditions. Another vital quality was the ability to live together in harmony for a long time without outside communication. Organizations – including NASA – that have studied how Shackleton survived find that even though his mission failed, every man survived the impossible odds because Shackleton had picked a good team and had made sure each member understood his role. Also, Shackleton knew how the rigors of Antarctic exploration would test the spirit of his men, so he was careful to look for character, and not just competence.

Technical qualifications were an asset, but he placed a greater emphasis on a positive attitude and a light-hearted nature. For example, when he interviewed Reginald James, who became the expedition's physicist, he asked whether James could sing. Alexander Macklin, a surgeon, won a place on the expedition when, in response to Shackleton's inquiry about why he wore glasses, Macklin replied, "many a wise face would look foolish without spectacles." Then there was the selection of his leaders. With over two dozen men to command, Shackleton understood the value of having loyal and strong leaders, which is why he chose Frank Wild as his second-in-command. Wild was a veteran of Antarctic exploration, who had more than proven his mettle and his compatibility with Shackleton on his 1907 expedition. Likewise, Thomas Crean, Shackleton's second officer, had proven his strength and discipline in his service with Shackleton on a 1901 expedition. So, how will Musk choose his interplanetary crew? He could do worse than apply Shackleton's guidelines:

Shackleton's 10 guidelines for choosing crewmembers.

1. Start with a solid crew you know from previous expeditions or who come recommended by those you trust.
2. Your second-in-command is the most important hire. Pick one who complements your management style, shows loyalty without being a yes-man, and can work with others.
3. Hire those who share your vision.
4. Weed out those who are not prepared to do mundane or unpopular jobs.
5. Go deeper than job experience and expertise. Ask questions that reveal a candidate's personality, values, and perspective on work and life.
6. Surround yourself with cheerful, optimistic people. Not only will they reward you with the loyalty and camaraderie vital for success, they will stick by you when times get tough.
7. Applicants hungriest for the job are apt to work hardest to keep it.
8. Hire those with the talents and expertise you lack.
9. Spell out clearly to new employees the duties and requirements of their jobs.
10. Help your crew do top-notch work.

That is because, in common with Shackleton, Amundsen and Co., Musk's group of astronauts will need to be prepared for changes of plan, contingencies, and the possibility that goals will not be achieved. It is this theme of contingencies and plans going awry that the media love to focus on. Working in these environments, the aforesaid doom and gloom merchants argue, will cause astronauts to lose their minds, turn on each other, and even come to blows. This is fantasy. Let us look at a real-life experience to prove the point.

In 1893, Fridtjof Nansen sailed to the Arctic in the *Fram*, a purpose-built, round-hulled ship designed to drift north through the sea ice. Nansen's theory was inspired by the voyage of the *Jeannette*, which foundered northeast of the New Siberian Islands and was found on the southwest coast of Greenland after having drifted across the Polar Sea. Nansen reckoned the Polar current's warm water was the reason for the movement of the ice. But, after more than a year in the ice, it became apparent that *Fram* would not reach the North Pole. So Nansen, accompanied by Hjalmar Johansen, continued north on foot when the *Fram* reached 84° 4′ North. It was a bold move, as it meant leaving the *Fram* not to return, and a return journey over drifting ice to the nearest known land, which was 800 kilometers south of the point where they started. Nansen and Johanssen started their journey on March 14, 1895 with three sledges, two kayaks and 28 dogs. On April 8, 1895, they reached 86° 14′ N, the highest latitude ever reached at that time. The men then turned around and started back, but they did not find the land they expected. On July 24, 1895, after using their kayaks to cross stretches of open water, they came across a series of islands where they built a hut of moss, stones, and walrus hides (Figure 8.14). Here they spent nine mostly dark months, spending up to 20 hours out of every 24 sleeping, waiting for the daylight of spring. They survived on walrus blubber and polar bear meat. In May 1896, Nansen and Johanssen decided to strike out for Spitsbergen. After travelling for a month, not knowing where they were, they were delivered from their endeavors through a chance meeting with Frederick George Jackson, who was leading the British Jackson-Harmsworth Expedition that was wintering on the island. Jackson informed them that they were on Franz Josef Land. Finally, Nansen and Johansen made it back to Vardø in the north of Norway.

Now, assuming the crew enjoys a problem-free deep space transit from Earth to Mars (highly unlikely, but you must be an optimist in the manned spaceflight arena), the next challenge awaiting them as they approach the Red Planet is performing the nail-biting EDL sequence which begins half an hour prior to touchdown. The EDL sequence includes a series of phases that must occur within a very narrow operational envelope, and most must be triggered autonomously, based on estimates of the where the spacecraft is relative to the ground and how fast it is traveling. Furthermore, each event must be executed flawlessly in the presence of potentially significant variability in Martian winds, atmospheric properties and surface topography. While several manned Mars mission architecture presentations exist, most conveniently gloss over the EDL details, thereby leading many people to conclude that landing humans on Mars should be easy.

Figure 8.14. The hut where Nansen and Johansen spent nine months. No Netflix, no streaming video, no North Face sleeping bags, no cell phones. Neither man had the advantage of years of astronaut training or spending months in analogs. Neither had written a psychological evaluation questionnaire. Together they proved that the human spirit is resilient and that the psychological aspect will be the least of mission planners worries when astronauts finally strike out for Mars. Public domain.

The reality, as we shall see, could not be more different, which is why seasoned scientists and engineers tasked with designing the EDL architecture regularly use phrases such as 'Six Minutes of Terror' to describe the anxiety evoked by sending a manned vehicle to Mars. These engineers and scientists, many of whom have been involved in sending robotic missions to the Red Planet, know that landing a spacecraft on the surface of Mars represents the most treacherous challenge of manned surface exploration. The harsh reality of the EDL problem is that, to date, 60 percent of all Mars missions have failed and many the ones to succeed have only done so by virtue of the spacecraft being small enough to allow them to reach the surface safely. Understandably, the problem of Mars EDL has generated much research, endless suggestions and volumes of articles but, to date, no one has solved the vexing problem of how a very large, manned vehicle travelling through the Martian atmosphere at Mach 5 can be decelerated to Mach 1 in 90 seconds and then re-orient itself from being a spacecraft to a lander and use engines to touch down. This is a quandary engineers refer to as the 'Supersonic Transition Problem' (STP) and this is how it may play out.

As the Starship approaches the Martian atmosphere, EDL software will be initiated, venting of the internal heat rejection system will be conducted and final trajectory correction maneuvers will be performed. The onboard computer will initialize attitude coordinates using a star scanner, which will inform the crew where Starship is relative to Mars by checking the positions of the stars. At this point the Starship will be just minutes from the interface with the Martian atmosphere. The onboard computer will continue to calculate and update its attitude and approach to the atmosphere, using Reaction Control System (RCS) thrusters to perform last-minute course adjustments. Shortly before entering the atmosphere, the autonomous guidance and navigation system will ensure the Starship's attitude is aligned at the precise angle required to perform the EDL maneuvers. As the Starship enters the Martian atmosphere at 8.5 kilometers per second, energy will be dissipated through heat transfer, causing the Starship to slow down. As the Starship descends through the atmosphere, heat built up around the vehicle will be dissipated through tiny holes in the windward side of the vehicle's stainless-steel shell, and once it has shed sufficient velocity, the vehicle will turn, fire its Raptor engines and land gently on the surface.

For those of you interested in the details, the author refers you to a paper published by the Aerospace Department of the University of Illinois, titled *Entry Trajectory Options for High Ballistic Coefficient Vehicles at Mars*. In a nutshell, what the researchers describe in their paper is how the Starship will use aerodynamic lift to steer through the atmosphere (Figure 8.15) before pulling up at the last moment and flying sideways through the thickest part of the atmosphere. The Starship would do this to dissipate most of its velocity before firing its descent engines and completing its powered landing. Hazard detection software and radar sensors will operate throughout the descent, providing the onboard computer with retargeting information based on altitude and horizontal velocity. The system will also generate new reference trajectories, depending on hazards detected and fuel expended. Powered descent will continue until approximately one meter above the surface, resulting in a vertical velocity at touchdown of two to three meters per second. After six or more months in deep space and less than ten minutes after entering the Martian atmosphere, the Starship will land on the surface of Mars and the crew will most likely breathe a collective sigh of relief. That will be because, while this sequence of events appears relatively straightforward on paper, the potential of the Martian environment to compromise one or more of these steps makes translating theory into reality a challenging and arduous task.

Spacecraft returning to Earth have the luxury of entering a thick atmosphere, capable of slowing a vehicle traveling at up to ten kilometers per second to less than Mach 1 while still 20 kilometers above the ground, just by using a heat shield. The rest of the return to Earth is achieved using drag, lift and parachutes. On Mars, however, the atmosphere is only one percent as dense as Earth's, meaning there is precious little air resistance to decelerate the Starship. In fact, the atmosphere of

Figure 8.15. The Starship steering its way through the upper reaches of the Martian atmosphere. Credit SpaceX.

Mars at its *thickest* is equivalent to Earth's atmosphere at approximately 35 kilometers altitude. Upon arrival at Mars, the Starship will be blistering along at more than eight kilometers per second. Due to the tenuous atmosphere, a heavy spacecraft, such as the Starship, may never attain the subsonic terminal descent velocity required to begin the propulsive maneuvers for landing. The Martian atmosphere further compounds the EDL problem because the time for the remaining EDL events is very small, even if the Starship finally manages to decelerate to subsonic velocity, leaving little or no margin for error. In fact, by the time a Starship is low enough to initiate its landing maneuver, it may already be too close to the surface to prepare for landing. As some robotic Mars explorers have learned the hard way, it is typically not the fall that kills you, but the landing.

Using current technology, we can expect the Starship to initiate its landing system one kilometer above the surface. With only 1,000 meters of altitude before touchdown, even the best landing system developed will be pressed for time to successfully complete the myriad hazard avoidance maneuvers expected during the Terminal Descent Phase (TDP). The problem of detecting hazards is compounded by the fact that orbital surveillance is only able to detect rocks greater than one meter in size, although thermal inertia can provide information concerning very rocky surfaces. As the Starship descends, radar altimetry, Doppler radar and touchdown targeting algorithms will provide information regarding possible hazards, but these systems can be 'spoofed' by features such as slopes and other surface shapes. Furthermore, horizontal errors may be induced when a wide beam from a Doppler radar measures surface-relative velocity over slopes. NASA designs many of its most mission-critical systems with double or triple levels of redundancy, and with good reason, since the failure of certain systems may result in Loss Of Mission (LOM) and / or Loss Of Crew (LOC). Due to the myriad ways LOM or LOC could occur during EDL, it is natural to assume that the Starship's EDL systems will be designed with multiple levels of redundancy, but this will not be the case. The reason is the very limited time span of the Mars EDL and the complexity of switching between systems in flight. As a result, most EDL subsystems must be designed to be non-redundant, a feature engineers refer to as *single string*.

Ideally, the Starship will land close to pre-positioned assets, but to do this it will need to be capable of exerting aerodynamic control over the atmospheric flight path, a feat yet to be realized in robotic missions. The Starship will also need to be capable of autonomously adjusting its flight within the Martian atmosphere, requiring real-time hypersonic guidance algorithms which have yet to be written, although there are companies developing software to resolve this capability gap. Assuming the capability gap is resolved, the guidance algorithms will be designed to work in conjunction with the approach navigation software and inertial navigation system to help the Starship perform a landing in proximity to the pre-positioned assets. During the descent to the surface, the guidance algorithm, which will be automatically linked to the RCS, will maneuver the Starship to accommodate off-nominal landing states and to account for any unusual atmospheric flight conditions such as surface winds, vertical gusts, and abnormally high or low atmospheric pressure.

The ability to develop, demonstrate and mature new EDL technologies will be key to realizing the goal of landing humans on Mars but, in developing these technologies, Musk's engineers must not only contend with the problems posed by the atmosphere but also ensure the survival of a deconditioned crew. The transition from hypersonic to touchdown speed in just a few minutes will require some ingenuity on the part of engineers to ensure deceleration forces do not disable the crew. Such a rapid transition from hypersonic to subsonic velocities also results in tremendous heat being generated, requiring adequate thermal protection strategies to be developed. In addition, since EDL events occur so quickly, systems must also be designed to function with the highest system reliability of any spacecraft ever developed. Because of all these EDL conundrums, it is not surprising that some engineers have described a landing on Mars as analogous to herding cats in a room with smoke and mirrors, where the floor is covered with apples and oranges (Figure 8.16).

Despite the challenges portrayed in the above scenario, there are tools in the works that may help Starship nail its landing. One way of ensuring the Starship enters the optimum entry corridor may depend on special software such as the Evolved Acceleration Guidance Logic for Entry (EAGLE). EAGLE is a

Figure 8.16. The EDL conundrum. Credit NASA.

hypersonic guidance algorithm developed for manned Mars missions by scientists at the University of California and at NASA's Jet Propulsion Laboratory (JPL). EAGLE simply consists of a trajectory planner and a trajectory tracker. During the entry phase, the trajectory tracker sends attitude commands to the autopilot, which calculates the correct angle of attack and the optimal aerodynamic lift force for the trajectory to be flown. Using information from accelerometers, aerodynamic acceleration is calculated and this information is sent to the trajectory tracker, enabling EAGLE to update the reference trajectory accordingly and keep the vehicle within the entry/flight corridor. To test the effectiveness of EAGLE, the software has been evaluated using the Mars atmospheric flight simulation in the

Dynamics Simulator for Entry, Descent and Surface landing (DSENDS) space-craft simulator at JPL. DSENDS is a high-fidelity spacecraft simulator that helps mission planners design EDL plans for vehicles landing on planets. In addition to simulating the actual flight of a vehicle through the Martian atmosphere, DSENDS is also able to integrate other computer models, such as those for thrusters, star trackers, gyros and accelerometers. The result is a simulator capable of providing mission planners with very accurate information about a specific EDL plan.

Based on promising results at JPL, it would seem the problem of ensuring the correct flight path is on track to being resolved. The next step is to ensure the crew is protected during the stresses of deceleration and the thermal loads encountered as the vehicle passes through the atmosphere.

9

Starship

"I know exactly what to do. I'm talking about moving there. We think you can come back but we're not sure. Your probability of dying on Mars is much higher than on Earth. It's gonna be hard. There's a good chance of death, going in a little can through deep space."

Elon Musk, in an interview with Mike Allen and Jim VandeHei of Axios for the HBO series.

Figure 9.0. Artist's depiction of the Starship. Credit SpaceX (public domain).

© Springer Nature Switzerland AG 2022
E. Seedhouse, *SpaceX*, Springer Praxis Books,
https://doi.org/10.1007/978-3-030-99181-4_9

People will probably die. Can you imagine the NASA Administrator being interviewed about the agency's Mars mission plans and telling reporters he/she expects some of the agency's astronauts will die? After the predictable uproar on Fox News, the agency would be shut down overnight. But Elon Musk is right. Mars, as we discussed in the previous chapter, is a risky proposition and it is almost certain that some of those venturing to the Red Planet will die. How many? Who knows, but that Starship had better pack a few body-bags just in case.

Auspiciously, that quote is from an interview with the Mars Messiah that occurred just a day before the scheduled landing of the Mars Insight spacecraft. Anyone who knows anything about Mars knows that the track record of landing safely on the planet is less than 50:50, so Musk's predictions of casualties, dire as they may sound, are probably not far off the mark. After all, this will be a survival mission, one that recalls Shackleton's epic expedition. Constant danger; safe return doubtful; hazardous journey; all virtually guaranteed. Why? Because Musk is planning on sending his crew of Musketeers to Mars before the decade is out. That is a bold plan even by the Mars Messiah's standards. But let us not forget who we are talking about here. This is someone who has built his reputation on pushing the envelope and then some. Musk would not be Musk if he did not make bold declarations, even if those declarations sometimes remind people of some uncomfortable truths, in this instance that some of those journeying to the Red Planet will not be making a return trip.

If you have read all the chapters of this book you will know by now that we are not dealing with a traditional visionary. Musk is someone who champions electric cars, hyper-loops… and colonizing Mars. This is someone who, against all expectations, made spaceflight considerably more affordable and accessible than it had been when government launchers dominated the industry. Just 12 years after founding his company, SpaceX successfully recovered a Falcon 9 first stage. Four years later, the Falcon Heavy made headline news with the recovery of two of its boosters. As this book is being written, Musk's Starship is being prepared for orbital test flights.

"People have told me that my timelines have historically been optimistic. So I am trying to recalibrate to some degree here. But I can tell what I know currently is the case is that we are building the first ship, the first Mars or interplanetary ship, right now, and I think we'll probably be able to do short flights, short sort of up-and-down flights probably in the first half of next year."

The Mars Messiah explaining his Mars mission timeline. Musk has a history of unrealistic timelines, but in an interview with *Time* magazine in December 2021, he stated he would be surprised if humans had not landed on Mars within five years. The first Starship (cargo) mission to the Red Planet is planned for 2024, followed two years later by a manned mission.

So, what exactly is the Starship (previously the Big Falcon Rocket – BFR – aka the Super Heavy, aka the Interplanetary Transport System – ITS)? While the name of the Starship may have changed several times, its mission has stayed the same: to ferry up to 100 people to Mars. On each flight, no less. The vehicle is a reusable system comprising a Super Heavy first stage and Starship second stage. Unusually for a launch vehicle, both stages are made from stainless steel (301 stainless steel alloy). Why? First, stainless steel is much cheaper than carbon composites (67 times cheaper), which was the material SpaceX initially planned to use to build the two stages. Second, steel is much easier to work with than carbon composites, which allows for speedier prototyping. Third, stainless steel performs better than carbon fiber when heated, and that is an important characteristic when the spacecraft travels through the atmosphere. So, temperature tolerance will be important as Starship (Table 9.1) makes its way through planetary atmospheres and through the freezing environment of deep space, but the thermal properties of stainless steel are still not sufficient to protect the vehicle from the scorching temperatures encountered during a trip through the Martian atmosphere. That is why SpaceX will cool the vehicle using a cryogenic liquid methane pump located underneath the Starship's skin. The stainless steel stages hold liquid oxygen and liquid methane which are channeled through the Super Heavy's cluster of Raptor engines to generate 16,000,000 pounds of thrust (about double the thrust of the Saturn V rocket).

Table 9.1: Starship Characteristics

Height	120 meters	**Super Heavy length**	70 meters
Diameter	9 meters	**Super Heavy propellant**	3,400 metric tons – liquid methane & liquid oxygen
Stages	2	**Super Heavy engines**	33 Raptors
Launch Site	Boca Chica & LC-39A	**Starship length**	50 meters
Payload to Low Earth Orbit	100 metric tons	**Starship propellant**	1,200 metric tons
Payload to Lunar Orbit	100 metric tons (with orbital refueling)	**Starship engines**	3 Raptors

Figure 9.1. Elon Musk contemplates the suboptimal landing of SN8. Credit Steve Jurvetson. Public domain.

The first test flight of a Starship test article was performed at the Boca Chica launch site by Starhopper on July 25, 2019. This was followed by a test of Starship test article SN8, which crashed on December 9, 2020 (Figure 9.1). After several more tests, SN15, which launched on May 5, 2021, became the first Starship article to land successfully. We will discuss some of these tests in more detail later in this chapter, but first let us take a closer look at the Super Heavy and Starship.

Powering the vehicle off the pad is the job of the Super Heavy booster, which has a dry mass of between 160 to 200 metric tons (the final design has yet to be decided). Providing the power is a cluster of 33 Raptors which will accelerate the spacecraft to Mach 9. Once all the fuel is expended, the Super Heavy will return to the launch site and land at the launch tower. During the descent to the launch pad, grid fins (which you can see at the top of Figure 9.2) will control the booster's pitch, aided by a Reaction Control System (RCS) powered by vented evaporated gas from the propellant tanks.

The Starship (Figure 9.3), which sits atop the Super Heavy, holds 1,200 metric tons of fuel. With its heatshield, the dry mass of the vehicle is 100 metric tons. At the base of the vehicle is a cluster of six Raptor engines, three configured for atmospheric pressure and three configured for vacuum operation. The liquid oxygen and liquid methane fuel tanks are separated by a dome, and above the fuel tanks is the payload bay which will hold crew cabins in the crewed variant (there

Figure 9.2. Super Heavy booster. Note the grid fins at the top of the booster. After separation from Starship, the booster fires gas thrusters to pitch over into a retrograde attitude in preparation for landing. Next is an orientation burn that fires the engines in the direction of travel to slow the booster down as it falls through the atmosphere. As the booster flies through the atmosphere, those grid fins help control the three-axis orientation of the booster as well as controlling down range distance. Once operational, the booster will be turned around within just hours of its return and be relaunched. Credit Lars Plougman. Public domain.

is also a cargo variant which we will discuss later). Once the Starship separates from the Super Heavy, the Starship's Raptor engines take over and propel the vehicle to orbit. Re-entry is enabled by a heat shield comprising ceramic hexagon tiles (Figure 9.4) that cover the windward side of Starship. These tiles, designed to deal with heat up to 1,350°C, are also designed with reusability in mind. Having said that, there is a Starship variant that does not require a heatshield because this variant will never re-enter an atmosphere. We are, of course, talking about the Starship Human Landing System (HLS) being developed for NASA's Artemis Program. Those SpaceX afficionados among you will recall the drama involved in selecting the Starship HLS because of the lawsuit filed by Blue Origin (see sidebar: *Lawsuit*)

Figure 9.3.   Starship SN16. Credit Lars Plougman. Public domain.

Lawsuit

In April 2021, NASA announced it had decided to give the contract for developing the HLS to SpaceX. In addition to SpaceX, the contract had been bid for by Dynetics and Blue Origin. On hearing the news, Blue Origin and Dynetics quickly did what Sierra Nevada Corporation (SNC) had done when it lost a NASA contract: they lodged a protest with the Government Accountability Office (GAO). The GAO reviewed and denied the protests in July 2021, prompting Blue Origin to file a lawsuit the following month, which resulted in SpaceX having to cease development of the HLS and in Jeff Bezos receiving hate mail from vociferous and agitated Musketeers. Ultimately, the lawsuit did not go the way Blue Origin hoped when a judge dismissed the case in November 2021, allowing the wheels to start turning once again on the HLS development. NASA issued the following statement:

"NASA will resume work with SpaceX under the Option A contract as soon as possible. In addition to this contract, NASA continues working with multiple American companies to bolster competition and commercial readiness for crewed transportation to the lunar surface. There will be forthcoming opportunities for companies to partner with NASA in establishing a long-term human presence at the moon under the agency's Artemis program, including a call in 2022 to U.S. industry for recurring crewed lunar landing services."

Musk, for his response, characteristically turned to Twitter, tweeting: *"You have been judged,"* alongside a screenshot from the film *'Dredd'*.

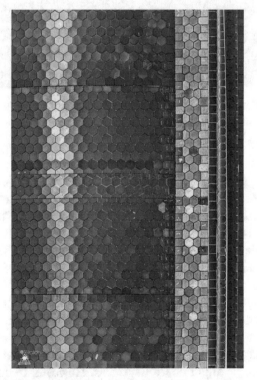

Figure 9.4. The Starship heat shield comprises ceramic hexagon tiles. Public domain

Figure 9.5. The Starship Human Lander System (HLS). Credit Steve Jurvetson/ SpaceX. Public domain.

The Starship HLS (Figure 9.5), which will feature a NASA docking system, solar panels and retro-propulsion, is the vehicle intended to return American astronauts to the surface of the Moon sometime in the mid-2020s. This is how that goal will be achieved. First, a Starship booster will launch an HLS into Low Earth Orbit (LEO) where waiting Starship tankers will refuel it. Then the HLS will boost itself into a Near Rectilinear Halo Orbit (NRHO), where it will meet up with a crewed Orion which will have been launched by NASA's Space Launch System (SLS). The Orion crew will then transfer to the HLS which will land on the Moon. The crew will stay for a few days and then the HLS will return the crew to Orion (Figure 9.6).

Because the Starship HLS will never enter an atmosphere, there is no heat shield and there are no flight control surfaces. However, in common with other Starship variants, the Starship HLS features six Raptor engines which will be used as primary propulsion and in other flight phases. In some mission designs there is talk of leaving the first Starship HLS vehicles on the lunar surface, where they could serve as a Moon base.

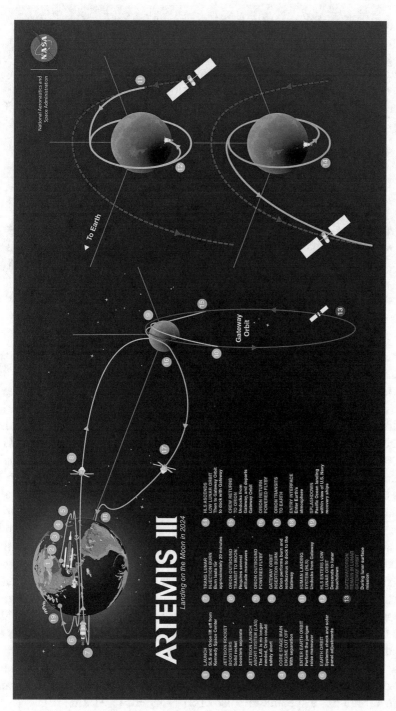

Figure 9.6. Depiction of NASA's Artemis III mission, which will be a crewed mission to the Moon sometime in 2024. Credit NASA.

Figure 9.7.   Starbase, Boca Chica. Credit Alexander Hatley. Public domain.

As this book is being written, preparations for Starship are well underway in Boca Chica, the SpaceX launch site in Texas. Except it is no longer just a launch site. It is a city named Starbase (Figure 9.7). If you happen to be one of the thousands of SpaceX faithful who have made their 'pilgrimage' to Starbase (the author himself knows of several people who have driven more than a thousand miles just to be able to say they have seen the 'holy' site), you will be familiar with the 'Welcome to Starbase' sign that sits in front of the 'Welcome to Boca Chica' sign. The site features a launch and landing pad and a launch tower, which stands 146 meters tall. It is an unusual launch tower, which sports a pair of mechanical arms with claws (and has been dubbed 'Mechzilla' by SpaceX fans) designed to catch boosters by closing around the booster. The system is a work in progress, and until 'Mechzilla' is operational, boosters will continue to land on drone ships or pads near the launch site. Not surprisingly, given that building and testing rockets is a noisy business, there have been more than a few protests from the locals, but these do not seem to have deterred Musk. On the subject of noise, we should now turn our attention to those aforementioned Raptor engines (Figures 9.8a and 9.8b).

The Raptor engine can generate 2.3 million MN of thrust. The engine is fueled by liquid methane and liquid oxygen via a full-staged combustion cycle (Figure 9.8b). The Raptor is available in two variants: the regular and the vacuum. The regular variant has a specific impulse of 330 seconds (3.2 km/sec), and the vacuum variant has a specific impulse of 378 seconds (3.71 km/sec).

Figure 9.8a. Raptor engine. Credit Brandon de Young. SpaceX. Public domain.

Another key feature being developed at Starbase are those tiles mentioned earlier. There has been much discussion on the various SpaceX blogs and in the media about the exotic technology of *transpiration cooling*, but it is unlikely that SpaceX will rely on this for their heatshield, or at least not the whole heatshield. SpaceX are not the market leader in all things crewed spaceflight because of technological leaps, nor are they where they are thanks to some secret exotic technology. No, SpaceX are in the position they are thanks primarily to streamlining and refining production of current technology and optimizing and refining current development and production processes.

Ever since its inception, SpaceX has favored simple and straightforward solutions over reinventing the wheel, following the 'best part is no part' philosophy. What better example of this, for instance, than the choice of stainless steel rather than carbon composites in the design of the Starship? When discussing the options of what type of heatshield to use on a spacecraft, most engineers will opt for an ablative system because these are simple, tried, tested and trusted systems, and if your goal is profit and expediting development of a revolutionary spacecraft, you do not want to be spending an inordinate amount of time and resources trying to

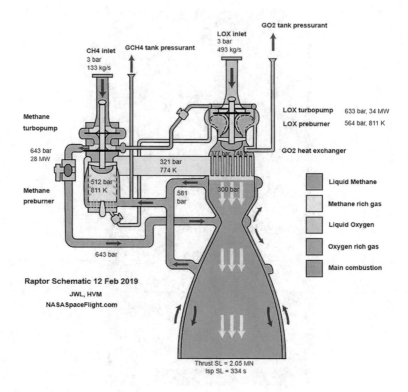

Figure 9.8b. Raptor engine combustion scheme. The way a full flow staged combustion cycle rocket engine works is as follows. Liquid methane and liquid oxygen are routed to turbopumps where the gases are pressurized, mixed and heated by pre-burners, one for the liquid methane and the other for the liquid oxygen. The high pressure generated in the turbopumps, and the high temperature generated in the pre-burners, causes the liquid methane and liquid oxygen to evaporate, which in turn causes the turbines and turbopumps to spin. This sequence is repeated until hot gas ignites inside the combustion chamber and the gas is directed through the nozzle to generate thrust. Of course, all this heat must be dissipated and that is achieved by circulating fuel around the fuel chamber. Credit NASA.

develop a new system. But the problem with ablative systems is that they are not reusable, and the Starship must have rapid reusability. So why not use tiles? After all, the Space Shuttle Orbiter had tiles.

Unfortunately, the Orbiter routinely returned from space with damage to its tiles, requiring months of repair and replacement work. That is something SpaceX cannot afford. The Starship must have a thermal protection system that is simple, that can be replaced with minimal inspection, and is able to withstand re-entry multiple times. So why has Musk opted for tiles, given all the headaches with the Orbiter's thermal protection system? Simply, the tiles SpaceX uses and the tiles used on the Orbiter are not the same. First, the Orbiter's tiles were much more fragile than the ones used on the Starship, because they were just a few inches away from the External Tank (ET) that was filled with cryogenic fuel and oxidizer. Because the ET operated at cryogenic temperatures it had to be covered in foam to minimize boil-off (it gets pretty hot in Florida), and the problem with that foam was that large pieces detached during launch, some of which hit and damaged the tiles covering the Orbiter. The Starship is bolted on top of a booster, so its thermal protection system will not be at risk of being hit by errant pieces of foam. Second, the Orbiter's tiles were made of rather soft and brittle materials (LI-900 silica), which meant that even minor impacts necessitated maintenance and replacement. In contrast, the Starship tiles are made of TUFROC, which is much more resilient. Third, the Orbiter's fuselage was made of aluminum which meant any breach in the tiles could be catastrophic, as evidenced in the demise of *Columbia* in February 2003. The Starship on the other hand, is constructed of stainless steel, which can absorb heat much more effectively. Fourth, the Orbiter's tiles were glued onto the fuselage and that glue represented a weak point (the Starship's tiles will be bolted on). All these differences combined add up to a huge contrast between the delicate thermal protection system of the Orbiter versus the more robust thermal protection system of the Starship. But there is more. Fifth, the Orbiter used thousands of uniquely shaped tiles whereas the Starship uses uniform hexagonal tiles, and finally, if SpaceX decides to use transpiration cooling, their thermal protection system may be even more robust.

But what exactly is a transpirational heat shield? Such a system relies on the heat capacity of vaporizing liquids to soak up thermal energy during re-entry. In the Starship, this would be achieved by designing a sandwich-type hull that is regeneratively cooled by liquid water. DLR, the German Space Agency, did some work on the concept in the 2010s using a porous material called Procelit 170 (P170) which demonstrated effective cooling capabilities. One downside to such a system is the amount of water required. Remember, the Starship will spend six minutes in Martian re-entry, which means the thermal protection system must be actively cooled with water for that amount of time. How much water will that require? If you read the research, you will see numbers in the 15-ton range. That is a lot of water and would impact significantly on the amount of payload SpaceX would be able to land on the surface of Mars. At the time of writing, it seems SpaceX is still considering transpirational cooling, but only for a small surface area of the Starship. We will have to wait and see.

So, assuming SpaceX can solve all its technology challenges – and that is not an easy assumption to make: after all we are not talking about Boeing or some government-sponsored aerospace dinosaur here – then how will Starship work operationally? Before we discuss operations, it is prudent to delve into the design history of what will surely become a revolutionary rocket. We will start with the first mention of what was to morph into the Starship. This event occurred in 2005, when our Mars Messiah mentioned a theoretical heavy lift launch vehicle known as the Big Falcon Rocket, or BFR. The BFR, also dubbed the Falcon XX, would be powered by Merlin engines and be capable of lifting 140 metric tons to LEO. Seven years later, the BFR was recalibrated as the Mars Colonial Transporter (MCT), which was then branded as a super heavy lift launch vehicle capable of lifting between 150 and 200 metric tons to LEO. To enable this behemoth to do its job, the MCT was to be powered by Raptors. Then, four years later when talking at the 2016 International Astronautical Congress (IAC) in Mexico, Musk came up with another name for his Mars transport, announcing plans to build the Interplanetary Transport System (ITS). Powered by no less than 42 Raptors, the ITS was planned to be capable of lifting a whopping 300 tons to LEO. Made from carbon composites, this titan of a launch vehicle would be cooled using the afore-mentioned transpiration process.

On the subject of the Raptors, by September 2017, the engines had tested 42 times and had been subject to 1,200 seconds of test firing, the longest of which was 100 seconds. That same year, SpaceX volunteered a few more details of their planned Mars missions, stating that the intent was to launch cargo missions in 2022. These cargo missions would be deployed to confirm water resources and to establish a power and life support infrastructure prior to the crewed flights, which were planned to launch in 2024. Fast track forward two more years and we find the first use of the names (Super Heavy booster and Starship) that are commonly used in the 2020s. On August 27, 2019, testing for what was to become the Starship began when a test article dubbed Starhopper reached 150 meters. This event was followed by the unveiling of Starship Mark 1 the following month and Mark 2 five months later, although neither flew.

At the beginning of 2020, SpaceX changed the naming of their test articles and switched to using serial numbers instead of Marks. So, Mark 3 became SN1 (Serial Number 1). While SN1 failed its cryogenic proof test in February 2020, the story for SN2 was a more successful one, but this test was followed by an unplanned depressurization of the SN3 article. Then, in May 2020, during one of SN4's static fire tests, a quick disconnect fuel line malfunction caused the article to explode. In August 2020, SN5 was ready for a hop to 150 meters, powered by one Raptor. This event marked the first time a full-scale test article had successfully completed a test flight. This flight was repeated by SN6. Next on the test agenda were static fire tests, which were conducted on SN8 during the months of October and November 2020. Four such tests were completed, with just one malfunction when the third test triggered an engine shutdown. Then, in December 2020, SN8 performed the first Starship flight and reached 12,500 meters, but the vehicle exploded on impact. Such an accident would stall a program for months, if not years, in a government-run schedule, but SpaceX hardly broke stride and conducted another flight just a few weeks later. Unfortunately, this test in February 2021 was also unsuccessful, when SN9 reached ten kilometers but suffered an engine malfunction. The following month saw SN10 repeat the profile of SN9 but the vehicle landed hard. More malfunctions followed when SN11 exploded in midair during its March 2021 flight. After that attempt, SpaceX decided to bypass SN12, SN13, and SN14 and fast-track to SN15, which eventually flew successfully in May 2021. As this book is being written, SpaceX have also decided to skip SN16, SN17, SN18, and SN19, planning instead to fly SN20 (Figure 9.9) on an orbital flight scheduled for the beginning of 2022.

Figure 9.9. SN20 gets some final touches before its orbital test. The first stage, or the Super Heavy, which is 63 meters high and nine meters wide, will have a gross liftoff mass of 3,065,000 kilograms. Constructed of stainless steel, it will be fueled by liquid oxygen and liquid methane and will be powered by 31 Raptor engines generating 61.8 million Newtons of thrust. The second stage, or Starship, features an integrated payload section that will serve to carry cargo or crew. It will be built in three versions: 1: a spaceship, with a pressurized volume of 1,000 m^3, to carry crew; 2: a tanker to carry fuel to LEO, and 3: a satellite delivery spacecraft. Credit Lars Plougman. Public domain.

So, with Starship up and running, how will SpaceX use it? One plan is to use Starship to replace the company's other rockets. Now you may be thinking that hardly sounds like a plan that makes much sense economically. After all, a comparably sized launch vehicle, the SLS, is forecast to cost at least two billion dollars for each launch, so surely the Starship must be an expensive proposition? Actually, the forecast cost of launching the Starship is just two *million* dollars, and that includes the propellant. This means that, given that Starship could carry about 400 Starlink satellites to orbit compared to the Falcon 9's capacity of 60, using Starship for routine SpaceX operations makes sense. Of course, the primary focus is Mars, but before SpaceX heads to the Red Planet, it will be helpful to launch one or two test missions to see how the whole kit and caboodle works in deep space. One such mission is the dearMoon project (Figure 9.10) funded by Japanese billionaire Yusaku Maezawa.

Figure 9.10. A violinist keeps the fabulously wealthy entertained during dearMoon's circumlunar mission. Credit SpaceX. Public domain.

The six-day dearMoon mission, announced in February 2017, will carry Maezawa and six to eight artist friends on a flyby mission of the Moon in 2023, on what will be the first lunar tourism mission ever. Why? One reason is that Maezawa hopes the mission will inspire the passengers to create something … well, inspiring. The plan is that such inspirational art will be exhibited following return to Earth to help promote peace. The dearMoon trajectory will be akin to that followed by Apollo 13 incidentally, and NASA plans to fly similar trajectories with its Artemis 1 and 2 launches. With Artemis 2, which will be the first crewed mission of the Artemis Program, not due to launch until 2025, SpaceX could very well beat NASA back to the Moon.

Assuming Maezawa's mission goes well, it will be all systems go for Musk's crewed Mars mission. The planning for this mission will no doubt be conducted at the same time as developing the aforementioned HLS, which brings us to a discussion of what the main elements of Musk's Mars mission will be. First, we have the Super Heavy reusable booster which we mentioned earlier. On top of that is the Starship, which will carry either cargo, propellant or crew. The plan for the Mars mission is that the Starship will launch to LEO and be refueled by another Starship carrying propellant. For the return trip, SpaceX will rely on In-Situ Resource Utilization (ISRU) to generate fuel. ISRU will also be used to generate oxygen from sub-surface water ice and from the atmosphere of Mars. The first flights, which will be cargo flights, may depart in 2026 and will deliver surface vehicles, food, life support infrastructure and mining equipment. If successful, the cargo flights may be followed by crew flights in 2028.

The crew missions, which will include temporary habitats (the Starships), will most likely land at around 40° latitude to support optimal solar production. This landing zone also happens to be reasonably close to sub-surface water ice deposits and is not too cold. Studies have shown that ISRU technology could mine water ice at a rate of one ton per day. As far as propellant is concerned, prototype units suggest one metric ton of oxygen/methane can be converted per 17-megawatt hours energy input from solar power. This means the ISRU units will be key to establishing a sustainable base, because the fuel load for a Starship is around 16 gigawatt-hours. Given that launch windows only open to Mars every 26 months, this means the ISRU units will have just over two years to generate the required fuel. To do this will require about one megawatt of continuous electric power, which means the solar panel ground-based array will need to cover more than 56,000 square meters. That sounds like a lot of solar panels, but the entire system should fit inside a Starship. The alternative will be to use a fission reactor, which weighs around 210 tons with associated power infrastructure, and will require at least two Starship launches. But before fission reactors can be launched, a rudimentary base must be established, and that will be achieved by the first crewed missions, manned by perhaps a dozen or so astronauts. These crews will build basic shelters, configure a basic propellant plant, set up the ISRU equipment, cobble together green living space habitats and set the stage for the settlement missions. How long will this take? Musk has always talked about having a self-sustaining city on Mars by 2050. That might be a little ambitious, but it depends on what size the city is. Eventually, if Musk's dreams are realized, colonization will be achieved, and he can make a start on the next goal: terraforming (Figure 9.11).

Figure 9.11.   Terraforming Mars. Credit Daein Ballard. SpaceX. Public domain.

Will Musk succeed? Probably. But perhaps not on the timeline he has set his company. After all, whenever you work on a project as unpredictable and intricate as crewed spaceflight you must expect the occasional failed test or launch. Anyone who follows SpaceX knows Musk is the eternal optimist when it comes to time-tables and target dates, and if anyone is going to land humans on Mars it will almost certainly be SpaceX. But landing and surviving are two different chal-lenges. Once on the surface, astronauts will face the slew of physiological hazards that may result in some not making the return journey. That could be the price paid for being a pioneer.

10

The Rise of SpaceX

The Route to Commercializing Low Earth Orbit

Figure 10.0. Crew Dragon approaches the International Space Station. Credit NASA.

In 2011, when the Space Shuttle *Atlantis* made its final flight, America marked the end of an era of publicly funded space exploration. But while the iconic winged spacecraft that had symbolized U.S. dominance in space for 30 years was retired

© Springer Nature Switzerland AG 2022   189
E. Seedhouse, *SpaceX*, Springer Praxis Books,
https://doi.org/10.1007/978-3-030-99181-4_10

to various museums, the journey did not stop with *Atlantis*. A year later, following the successful flight of Dragon, SpaceX picked up the baton and is now carrying NASA's legacy forward. While SpaceX benefitted from sizeable chunks of NASA funding, Elon Musk still spent a significant share of his private fortune developing the Falcon family of launchers, the Dragons, and seeing his company through some tight spots along the way. The successes of the Falcon 9 and the Dragons undoubtedly represent a watermark in spaceflight history, ushering in the beginning of a new era when private companies, rather than governments, will challenge the 'final frontier'.

Future crewed Mars missions aside, it is worth noting that SpaceX has achieved so much already. It commands the commercial space industry, an industry previously dominated by nation states. It has bypassed traditional contractors such as Boeing and Lockheed Martin. It has also won just about every launch contract worth winning, whether that be the use of the Falcon Heavy to launch military satellites or the use of the Starship to land astronauts on the Moon before the end of the 2020s. On the media front, SpaceX has upstaged every space newsworthy event for years. Remember in July 2021, when Sir Richard Branson took his joyride to the edge of space and when Jeff Bezos made his trip on New Shepard? Those events were completely eclipsed by the SpaceX Inspiration4 mission which carried four passengers more than five times higher for a three-day journey around the Earth. For the SpaceX faithful, suborbital jaunts are so… yesterday. They want to get to Mars now. Then there are the hundreds of thousands of orders for the SpaceX Starlink broadband network, and the likelihood of astronauts who are scheduled to fly on Boeing's Starliner being switched to fly on the Crew Dragon. Remember Boeing? Back in the day it was a juggernaut aerospace company that practically defined an era of aerospace. Now, it has been usurped by what was once a scrappy start-up that was initially written off by many in the industry. How wrong those naysayers were. Because on just about *every* noteworthy metric, SpaceX leads the way by a distance.

Take the Falcon 9 (Figure 10.1). This revolutionary vehicle has significantly brought down the cost of carrying payload to space and has in many ways played a significant role in Musk's dream of reaching Mars. When measured in the metrics of performance, cost and reliability, the vehicle is quite simply the most successful rocket ever. In the first half of 2021, the closest launch competitor to SpaceX in the global launch market was China, which managed to launch almost as many rockets as SpaceX. But SpaceX launched almost three times as much weight into space as China. Not only that, but none of those Chinese rockets had boosters that flew back to the launch site. No pushing of disruptive technologies (3-D printing engines or reusing boosters, for example) into the mainstream for the Chinese. We should also not forget that SpaceX assumed full development risk for almost all these technologies, which had the added secondary benefit of the

Figure 10.1. In the business of making launch vehicles cheaper and space travel more affordable, mainly thanks to the Falcon 9, the SpaceX workhorse. Credit SpaceX, public domain.

company becoming much more financially disciplined. This in turn made SpaceX an even more formidable competitor in the aerospace industry.

Take the development cost of the Falcon 9 as an example of this. NASA reckons the $400 million SpaceX spent developing the Falcon 9 was a mere ten percent of the cost of a rocket developed using traditional government contracting. Then there is the funding of new ventures such as the Mars Messiah's crewed Mars mission. When most businesses want to fund new ventures they must generate cash from existing businesses, but the money required for the Mars venture has been raised from the private market thanks to the success of Starlink, which beat OneWeb and Kuiper to market. The sheer dominance of SpaceX has left its rivals frustrated and exasperated, as evidenced by Blue Origin's ultimately futile legal maneuver against SpaceX for the Moon landing award. That now infamous lawsuit complained that by choosing SpaceX, the industry would be deprived of a market for Blue Origin's New Glenn rocket. As anyone with even a flimsiest understanding of the aerospace industry knew, the lawsuit never stood a chance of succeeding for the simple reason that New Glenn (which had cost more than $2.5 billion by mid-2021) was not even close to reaching the launch pad, never mind being launched.

"Before SpaceX we only really had the ULA, so we're in a better position than we were."

Phil McAlister, director of NASA's commercial space flight division.

But despite all these successes, there are still those who moan. Warnings of the risks of a monopoly are amongst the most common. Others warn that a vertically integrated rocket company will weaken supply chains. The reality, unsurprisingly, is that most new rocket companies, having witnessed the success of SpaceX first-hand, have adopted a similar model. As for the supply chain, that served a *political* purpose, not a commercial one. By having hundreds of suppliers dotted around the country, lots of politicians have been able to claim success by winning government contracts. But with SpaceX, the effect of a vertically integrated company has been to lower launch costs dramatically and improve access to space. This has brought an *increase in demand*, so the supply chain warning is meritless. On the subject of launch costs, it is worth emphasizing by how much these figures have been reduced. Before SpaceX came along, the price of getting a kilogram of payload into orbit was $15,000 or more. With SpaceX in the game that kilogram cost was reduced to just $5,000, and with Starship around the corner that cost could fall even further, to just $500. Lockheed Martin and ULA have little chance of competing.

At those costs it makes little sense in keeping the Falcon 9 around, so it is not surprising to hear that Musk intends to retire the old workhorse once the Starship is up and running. Along the way of developing and testing his Starship, Musk has provided key lessons for the rest of the aerospace industry. One of these is to take proven rocket designs, simplify them and then streamline them as much as possible to build them quickly and cheaply. Since the SpaceX chief rocket designer does not have to worry about distributing jobs geographically, it is a strategy that has paid off. One of the problems NASA has always had – and still does – when it comes to keeping costs down is that the space program has always been seen by politicians as a way of creating jobs, especially in poorer regions of the United States. That is one of the reasons why the NASA launch facilities are in Florida, but Mission Control is in Texas. It also explains why there are space centers in places like Alabama. With such wide a distribution of people and facilities, it is hardly surprising that it cost north of a billion dollars every time the Space Shuttle was launched. SpaceX, on the other hand, uses as few people as possible, in as central a location as possible, to build rockets in as few steps as possible. Having said that, being able to draw on NASA's expertise has certainly paid off for SpaceX, as is evident in the Dragon's design. The shape of the capsule is derived from that of Apollo and utilizes similar, slightly steerable aerodynamic characteristics during re-entry. The Dragon's heat shield also shares Apollo heritage, except that the SpaceX version is reusable whereas the ablative Apollo version was not. The capsule's escape system is also an Apollo legacy item although this, in common with the heat shield, is also being improved.

It is also worth emphasizing that previous point of SpaceX not having blazed its trail of glory alone. Accessing NASA's vast experience has not only allowed SpaceX to save time but also to develop components cheaply, although saving money is only part of Elon Musk's philosophy. Musk believes that to truly make spaceflight cheap you need to make the flight components reliable, and that means *reusable*. The logic is straightforward: simple systems become more reliable and reusable, reliable systems become simpler and reusable, and reusable systems become simpler and more reliable. One of the most frustrating aspects of spaceflight over the last few decades has been that it is such a wasteful enterprise. Thousands of person-hours are spent building these wonderful machines, which get used once and then either end up burning up in the atmosphere, sinking to the bottom of the ocean or gathering dust in a museum. The only partially reusable system that had any success was the Shuttle, but this spacecraft was a rats-nest of labyrinthine complexity. For those who believe in the value of a permanently expanding human future in space, like Musk and the SpaceX Musketeers, the realization and implementation of a fully reusable launch system is, and always has been, the key to achieving that dream. It is a dream that has been a long time coming and SpaceX is still a few years from achieving it, but with no previous companies playing the game of pursuing reusability, you must give credit to Musk for providing a credible path to accomplishing it.

Still, the company has its critics. Some speculate that SpaceX will find it difficult to match the quality control that is second nature for an organization like NASA. Others insist a crewed mission to Mars is a non-starter because there is simply no reason to go. What we do know is that, thanks to SpaceX, NASA now has a vehicle capable of serving as a lifeboat at the ISS in the event of an emergency necessitating evacuation. More importantly, thanks to the successes of Crew Dragon, the agency is no longer held hostage to the financial whims of the Russians and an aging spacecraft that has erratic and unpredictable re-entry reliability. Furthermore, Crew Dragon can support seven crew members, more than twice the complement the Soyuz can carry, which means the crew can be increased without safety concerns. Having the Dragon docked at the ISS also means the station has a functional ambulance capability which could return a sick or injured astronaut to Earth without the entire crew having to abandon the station. Lastly, the agency now has a way of getting astronauts to the ISS again after the retirement of the Shuttle, a capability it did not have in 2011 when the Russians experienced a launch failure, prompting NASA to consider abandoning the station. The Russian launch failure was a pivotal event in the history of the ISS because it underscored the reliance the U.S. had upon the Soyuz, so it is worth revisiting.

On August 24, 2011, an unmanned Russian cargo ship carrying supplies for astronauts on the ISS suffered a major malfunction after launch and crashed. The Progress 44 cargo ship had blasted off atop a Soyuz U rocket from Baikonur Cosmodrome in Kazakhstan, but less than six minutes into flight, shortly after the third stage was ignited, the vehicle commanded an engine shutdown. The capsule had been packed with almost three tons of food, fuel and supplies for the six astronauts on board the ISS. The crash was unusual because the Progress vehicle had a long track record of reliability, having serviced the ISS since the first crew occupied the station in 2001. No-one was more shocked than the Russians, whose Soyuz workhorse had logged 745 successful launches against just 21 failures since the vehicle came into service in 1966. Had a crew been onboard they probably would have been able to execute a successful abort, but given the small margin of error for future flights the Soyuz vehicle was grounded until further notice.

While there was no risk to the crew, who had plenty of supplies, the crash was a setback to the planned crew rotation, because a fresh crew of three was supposed to have been launched on September 22 to relieve half of the station occupants. If that flight did not happen, the incumbent crew could only stay onboard for so long. While two Soyuz spacecraft were docked at the station, serving as lifeboats in extremis, safety rules prohibited the vehicles from remaining in space for more than 200 days because their batteries could lose power (this is not a problem with Crew Dragon's solar array), and corrosive thruster fuel could degrade rubberized seals. With the clock expiring in September for one of the capsules and early December for the other, the astronauts were faced with the bleak prospect of switching off the lights, shutting the hatch and vacating the station without the arrival of a new crew. Fortunately, the 2011 incident was resolved, and NASA breathed a sigh of relief. But that was not the last problem. On October 11, 2018, Soyuz MS-10 (the 139th flight of the booster) was due to ferry two members of Expedition 57 to the ISS. A few minutes after liftoff, a booster failure triggered deployment of the launch escape system, pulling the capsule away from the rocket. Crewmembers Aleksey Ovchinin and NASA astronaut Nick Hague were recovered without suffering any injuries.

Those two incidents said a lot about NASA's (and the U.S. Government's) long-term planning skills, because the agency had known for years that the Shuttles were living on borrowed time. Yet here they were with only one transport provider and no redundancy. In 2004, the Bush Administration had launched a bold return-to-the-Moon program, instructing NASA to build two new launch vehicles, a new crew capsule *and* a lunar lander. That was the Constellation program, but it did not last long because NASA and the Obama Administration scrapped it in 2010, deciding to outsource LEO operations to the private sector. That decision left a space access gap that was not guaranteed to be filled by a new NASA vehicle anytime soon… until SpaceX came along.

Sixty years ago, Wernher von Braun, of Saturn V fame, was the voice of the US Space Program. A brilliant man, it was von Braun's vision and engineering expertise that propelled the U.S. to the Moon. Without von Braun, many doubt Kennedy's dream of reaching the Moon before the end of the decade would have been achieved, because visionaries like von Braun keep such projects from dying on the vine. Fortunately, in Elon Musk, the commercial spaceflight industry has its von Braun, capable of creating rockets and daring to dream big, so it was only fitting that Musk was awarded the Wernher Von Braun Memorial Award[1] in 2009.

Since the Shuttle *Atlantis* made its final landing, there have been innumerable op-ed pieces griping and grousing that NASA's temporary pause in the United States government's manned space flight capability program has meant that America has somehow abandoned space and that the U.S. has taken a step back from national greatness that embodied the Apollo era. But with the successes of SpaceX, it is hard to argue that the country has abandoned space and, with the company's launch cost reduction program, the cost of space access is going down. Let us also not forget that von Braun's ultimate goal was a crewed mission to Mars, a vision shared by the Mars Messiah.

On that note, as SpaceX takes over more and more of American advancement in space, some observers are asking the obvious questions: Can the task of space exploration be left to private industries, and are private industries even capable of tackling space development safely and efficiently without government oversight and funding? Having witnessed the recent successes of SpaceX, many observers would answer in the affirmative, but before commercial space companies start planning for missions to Mars, there needs to be some evidence that SpaceX and its competitors are capable. Now, obviously we cannot point to past examples of private industries developing space, but there are precedents in the field of exploration. For example, the University of Washington and Emory University completed a study that compared public and private expeditions to the North Pole between 1818 and 1909. Back in those days, Arctic and Antarctic exploration were funded either by government largesse or from the pockets of wealthy financiers. The study is an interesting one, because Arctic exploration and space exploration share many common features. First, there are the inhospitable conditions that test even the most advanced technologies. Then there are the myriad unknown and unforeseeable dangers. Thanks to the meticulous diaries and detailed ship logs kept by polar explorers, researchers were able to assess variables such as crew size, vessel tonnage, past experience of captains, and the number of deaths on the expedition. After a rigorous assessment of all those factors they came up with the following conclusion:

---

[1] The award is presented in odd-numbered years to recognize excellence in the management of and leadership in a significant space-related project. The award was originally proposed in 1992 by National Space Society Awards Committee member Frederick I. Ordway III, a close associate of von Braun.

"Most major Arctic discoveries were made by private expeditions. Most tragedies were publicly funded. Public expeditions were better funded than their private counterparts, yet lost more ships, experienced poorer crew health, and had more men die. Public expeditions' poor performance is not attributable to differences in objectives, available technologies, or country of origin. Rather, it reflects a tendency toward poor leadership structures, slow adaptation to new information, and perverse incentives."[2]

So, what does this have to do with the question of whether or not commercial companies can tackle space exploration? According to the study's findings, not only would the private sector be capable, but it would probably do better than publicly funded competitors. The study concluded that one of the problems with publicly funded expeditions is that they are slow to adapt to new information. Such a statement is still true today, as evidenced by the fact that NASA used the Shuttles for 30 years and then, instead of improving them, put them in museums. The same argument can be applied to the use of the Soyuz – 1960s technology that is still being used five decades later. Another liability of a public funded venture, as the study points out, is the problem of incentives. Publicly funded expeditions have a set budget that pays people the same amount regardless of the outcome – whether you do an outstanding job or one that is decidedly sub-par, you get paid the same. Not so with a private company, which is positioned to make profits if the expedition goes well. In this case, better performance means greater gains, which is exactly how incentives are supposed to work.

So, will it be SpaceX leading the way? There is no question that, at the time of writing, the company is on a roll, but there are still questions that need to be answered. One of these is whether the company can ramp up and maintain steady, reliable operations *and* low prices. Given the number of Falcon 9 and Dragon vehicles flying, combined with the aggressive testing that the Falcon rockets are subjected to before launch, and the control SpaceX has over its own component supply, it is difficult to see the vehicles having a bad day. As Bill Gerstenmaier, NASA's Human Spaceflight Administrator, stated in one of the many press interviews; "there is none better" than the SpaceX team. Another question is whether the company's competitors will deliver. That is difficult to say. At the time of writing, United Launch Alliance is facing spiraling costs and companies such as Orbital Sciences Corporation are several years behind. China, the lowest cost foreign competition to SpaceX, has no chance of competing with the company's pricing and, as far as foreign alternatives go, the only reliable launcher is the decidedly pricey European Ariane. Even when it comes to emergent launch capabilities, SpaceX either has no competition (in the heavy lift arena) or has the edge on their competition (reusability). So, assuming SpaceX does not suffer a major setback over the next few years, and assuming another strong competitor does not emerge, most routes to orbit and beyond may go through SpaceX by sometime in the 2020s.

---

[2] Karpoff., J. Public versus Private Initiative in Arctic Exploration: The Effects of Incentives and Organizational Structure. Journal of Political Economy. Volume 109, Number 1. February 2001.

Appendix I

Elon Musk Testimonies

ELON MUSK: PRESIDENTIAL COMMISSION SPEECH

As members of the Commission are aware, the cost and reliability of access to space have barely changed since the Apollo era over three decades ago. Yet in virtually every other field of technology, we have made great strides in reducing cost and increasing capability, often in ways we did not dream existed. We have improved computing costs by a factor 10,000 or more, decoded the human genome and built the Internet. The exception to this wave of development has been space launch, but why?

My best guess at the origin of the problem relates to a breakdown of what the economist Schumpeter called "creative destruction". He postulated that the way an industry improves is that new companies enter a market with a lower price or superior product. This forces the whole market to improve. Looking at orbital launch vehicles, we see a situation where there has been no successful new entrant in four decades, apart from one firm established in the late 1980s. Moreover, there has actually never been a truly commercial development anywhere in the world that reached orbit.

To address this problem, we must create a fertile environment for new space access companies that brings to bear the same free market forces that have made our country the greatest economic power in the world. If we can create such an environment, I expect that progress in space launch costs and capability will be no less dramatic than in other technology sectors.

If you doubt that we can possibly see such progress in space access, please reflect for a moment that the Internet, originally a DARPA funded project, showed negligible growth for over two decades until private enterprise entered the picture and made it accessible to the general public. At that point, growth accelerated by more than a factor of ten. We saw Internet traffic grow by more in a few years than the sum of all growth in the prior two decades.

© Springer Nature Switzerland AG 2022
E. Seedhouse, *SpaceX*, Springer Praxis Books,
https://doi.org/10.1007/978-3-030-99181-4

We are at a crucial turning point today. The vision outlined by the President is absolutely achievable within the current NASA budget and schedule, *but only by making use of new entrepreneurial companies along with the incumbents*. It cannot be achieved at all if we simply follow the old paths, which have led us to one cancelled program after another since the Space Shuttle.

What strategies are key to achieving the President's vision?

1. Increase and Extend the Use of Prizes.
Offering substantial prizes for achievement in space could pay enormous dividends. We are beginning to see how powerful this can be by observing the recent DARPA Grand Challenge and the X Prize. History is replete with examples of prizes spurring great achievements, such as the Orteig Prize, famously won by Charles Lindbergh, and the Longitude prize for ocean navigation.

Few things stoke the fires of creativity and ingenuity more than competing for a prize in fair and open competition. The result is an efficient Darwinian exercise with the subjectivity and error of proposal evaluation removed. The best means of solving the problem will be found and that solution may be in a way and from a company that no-one ever expected.

One interesting option might be to parallel every NASA contract award with a prize valued at one tenth of the contract amount. If another company achieves all of the contract goals first, they receive the prize and the main contract is cancelled. At minimum, it will serve as competitive spur even after contract award.

We should strongly support and extend the proposed Centennial Prizes put forward in the recent NASA budget. No dollar spent on space research will yield greater value for the American people than those prizes.

2. Support new entrants in space launch.
The most fundamental barrier to human exploration beyond low Earth orbit, and hence meeting the President's vision, is the cost of access to space. Here it should be noted that the cost of launch also drives the cost of spacecraft. If you are paying $5,000 per pound to put something in space, you will naturally pay up to $5,000 per pound to save weight on your spacecraft, creating a vicious cycle of cost inflation.

This problem of affordability dwarfs all others. If we do not set ourselves on the track of solving it with a constantly improving price per pound to orbit, in effect a Moore's law of space, neither the average American nor their great-grandchildren will ever see another planet. We will be forever confined to Earth and may never come to understand the true nature and wonder of the Universe.

It was precisely for this reason that I established SpaceX and set as our goal improving the cost as well as the reliability of access to space. Our first offering, called Falcon, will be the only semi-reusable orbital rocket apart from the Space Shuttle. Initially, we will deliver cargo to orbit in the form of satellites, however we believe strongly in the long term market for commercial human transportation.

As a starting point for improving the affordability problem, the Falcon is only one fifth the NASA list price of our U.S. competitors. Moreover, we expect to decrease our prices in real, if not absolute terms every year, and will be announcing a price decrease in our Falcon I vehicle shortly.

New companies might also provide reliability levels more comparable with airline transportation. In the case of SpaceX, we believe that our second generation vehicle in particular, the Falcon V, will provide a factor of ten improvement in propulsion reliability. Falcon V will be the first US launch vehicle since the Saturn V Moon rocket that can complete its mission even if an engine fails in flight – like almost all commercial aircraft. In fact, Saturn V, which had a flawless flight record, was able to complete its mission on two occasions only because it had engine out redundancy.

My thanks for the opportunity to come before you today, and I look forward to answering any questions that you may have.

ELON MUSK - SENATE TESTIMONY MAY 5, 2004

Mr. Chairman and Members of the Committee, thank you for inviting me to testify today on the future of Space Launch Vehicles and what role the private sector might play.

The past few decades have been a dark age for development of a new human space transportation system. One multi-billion dollar Government program after another has failed. In fact, they have failed even to reach the launch pad, let alone get to space. Those in the space industry, including some of my panel members, have felt the pain first hand. The public, whose hard earned money has gone to fund these developments, has felt it indirectly.

The reaction of the public has been to care less and less about space, an apathy not intrinsic to a nation of explorers, but born of poor progress, of being disappointed time and again. When America landed on the Moon, I believe we made a promise and gave people a dream. It seemed then that, given the normal course of technological evolution, someone who was not a billionaire, not an astronaut made of "The Right Stuff", but just a normal person, might one day see Earth from space. That dream is nothing but broken disappointment today. If we do not now take action different from the past, it will remain that way.

What strategies are critical to the future of space launch vehicles?

1. Increase and Extend the Use of Prizes.
This is a point whose importance cannot be overstated. If I can emphasize, underscore and highlight one strategy for Congress, it is to offer prizes of meaningful scale and scope. This is a proposition where the American taxpayer cannot lose. Unlike standard contracting, where failure is often perversely rewarded with more money, failure to win a prize costs us nothing.

Offering substantial prizes for achievement in space could pay enormous dividends. We are beginning to see how powerful this can be by observing the X Prize, a prize for suborbital human transportation, which is on the verge of being won. It is a very effective use of money, as vastly more than the $10 million prize is being spent by the dozens of teams that hope to win. At least as important, however, is the spirit and vigor it has injected into the space industry and the public at large. It is currently the sole ember of hope that one day they too may travel to space.

Beyond space, as the Committee is no doubt aware, history is replete with examples of prizes spurring great achievements, such as the Orteig Prize for crossing the Atlantic nonstop by plane and the Longitude prize for ocean navigation.

Few things stoke the fires of creativity and ingenuity more than competing for a prize in fair and open competition. The result is an efficient Darwinian exercise with the subjectivity and error of proposal evaluation removed. The best means of solving the problem will be found and that solution may be in a way and from a company that no-one ever expected.

One interesting option might be to parallel every major NASA contract award with a prize valued at one tenth of the contract amount. If another company achieves all of the contract goals first, they receive the prize and the main contract is cancelled. At minimum, it will serve as competitive spur for cost plus contractors.

Some people believe that no serious company would pursue a prize. This is simply beside the point: if a prize is not won, it costs us nothing. Put prizes out there, make them of a meaningful size, and many companies will vie to win, particularly if there are a series of prizes of successively greater difficulty and value.

I recommend strongly supporting and actually substantially expanding upon the proposed Centennial Prizes put forward in the recent NASA budget. No dollar spent on space research will yield greater value for the American people than those prizes.

2. Rigorously Examine How Any Proposed New Vehicle Will Improve the Cost of Access to Space.

The obvious barrier to human exploration beyond low Earth orbit is the cost of access to space. This problem of affordability dwarfs all others. If we do not set ourselves on the track of solving it with a constantly improving price per pound to orbit, in effect a Moore's law of space, neither the average American nor their great-great-grandchildren will ever see another planet. We will be forever confined to Earth and may never come to understand the true nature and wonder of the Universe. So it is critical that we thoroughly examine the probable cost of alternatives to replacing the Shuttle before embarking upon a new development. The Shuttle today costs about a factor of ten more per flight than originally projected and we don't want to be in a similar situation with its replacement.

In fact, it was precisely to improve the cost and reliability of access to space, initially for satellites and later for humans, that I established SpaceX (although some of my friends still think the real goal was to turn a large fortune into a small one). Our first offering, called Falcon I, will be the world's only semi-reusable orbital rocket apart from the Space Shuttle. Although Falcon I is a light class launch vehicle, we have already announced and sold the first flight of Falcon V, our medium class rocket. Long term plans call for development of a heavy lift product and even a super-heavy, if there is customer demand. We expect that each size increase would result in a meaningful decrease in cost per pound to orbit. For example, dollar cost per pound to orbit dropped from $4,000 to $1,300 between Falcon I and Falcon V. Ultimately, I believe $500 per pound or less is very achievable.

3. Ensure Fairness in Contracting.

It is critical that the Government acts and is perceived to act fairly in its award of contracts. Failure to do so will have an extremely negative effect, not just on the particular company treated unfairly, but on all private capital considering entering the space launch business.

SpaceX has directly experienced this problem with the contract recently offered to Kistler Aerospace by NASA and it is worth drilling into this as a case example. Before going further, let me make clear that I and the rest of SpaceX have a high regard for NASA as a whole and have many friends and supporters within the organization. Although we are against this particular contract and believe it does not support a healthy future for American space exploration, this should be viewed as an isolated difference of opinion. As mentioned earlier, for example, we are very much in favor of the NASA Centennial Prize initiative.

For background, the approximately quarter billion dollars involved in the Kistler contract would be awarded primarily for flight demonstrations and technology showing the potential to resupply the Space Station and possibly for transportation of astronauts.

That all sounds well and good. The reason SpaceX is opposing the contract and asking the General Accounting Office to put this under the microscope is that it was awarded on a sole source, uncompeted basis to Kistler instead of undergoing a full, fair and open competition. SpaceX and other companies (Lockheed and Spacehab also raised objections) should have, but were denied, the opportunity to compete on a level playing field to best serve the American taxpayer. Please note that this is a case where SpaceX is only asking for a fair shot to meet the objectives, not demanding to win the contract.

The sole source award to Kistler is mystifying given that the company has been bankrupt since July of last year, demonstrating less than stellar business execution (if a pun is permitted). Moreover, Kistler intends to launch from Australia using all-Russian engines, raising some question as to why this warrants expenditure of American tax dollars.

Appendix II

SpaceX Commercial Resupply Services (CRS) Missions

Mission	Capsule No.	Launch date	Remarks	Time at ISS (dd:hh:mm)	Outcome
SpX-C1	C101	2010 Dec 8	First Dragon mission, second Falcon 9 launch. Mission tested orbital maneuvering and reentry of the Dragon.	N/A	Success
SpX-C2+	C102	2012 May 22	First Dragon mission with complete spacecraft, first rendezvous mission, first berthing with ISS.	5:17:47	Success
CRS-1	C103	2012 Oct 8	First CRS mission for NASA, first non-demo mission. Falcon 9 rocket suffered a partial engine failure during launch but delivered Dragon to orbit. Secondary payload did not reach its correct orbit.	17:22:16	Success; launch anomaly
CRS-2	C104	2013 Mar 1	First launch of Dragon using trunk section to carry cargo. Launch was successful, but anomalies occurred with thrusters after liftoff. Thruster function later restored and orbit corrections made, but spacecraft's rendezvous with ISS was delayed from Mar 2 until Mar 3, 2013, when it was berthed with Harmony. Dragon splashed down in Pacific on Mar 26, 2013	22:18:14	Success; spacecraft anomaly
CRS-3	C105	2014 Apr 18	First launch of redesigned Dragon: same outer mold line with avionics and cargo racks redesigned to supply more	27:21:49	Success

© Springer Nature Switzerland AG 2022
E. Seedhouse, *SpaceX*, Springer Praxis Books,
https://doi.org/10.1007/978-3-030-99181-4

Mission	Capsule No.	Launch date	Remarks	Time at ISS (dd:hh:mm)	Outcome
			power to powered cargo devices, including additional cargo freezers for transporting critical science payloads.		
CRS-4	C106	2014 Sep 21	First launch of Dragon with living payload, in the form of 20 mice which were part of a NASA experiment to study the physiological effects of long-duration spaceflight.	31:22:41	Success
CRS-5	C107	2015 Jan 10	Cargo manifest change due to Cygnus CRS Orb-3 launch failure.	29:03:17	Success
CRS-6	C108	2015 Apr 14	The Dragon splashed down in the Pacific on May 21, 2015.	33:20:00	Success
CRS-7	C109	2015 Jun 28	Mission was supposed to deliver first of two International Docking Adapters (IDA) to modify Russian docking ports to newer international standard. Payload lost due to in-flight explosion of carrier rocket. Dragon survived the blast; it could have deployed parachutes and performed a splashdown, but software did not take this situation into account.	N/A	Failure
CRS-8	C110	2016 Apr 8	Delivered Bigelow Expandable Activity Module (BEAM) module in unpressurized trunk. First stage landed for first time successfully on sea barge. A month later, Dragon was recovered, carrying downmass	30:21:03	Success

Mission	Capsule No.	Launch date	Remarks	Time at ISS (dd:hh:mm)	Outcome
			containing Scott Kelly's biological samples.		
CRS-9	C111	2016 Jul 18	Delivered IDA-2 to modify the ISS docking port Pressurized Mating Adapter-2 (PMA-2) for Commercial Crew spacecraft. Longest time a Dragon was in space.	36:06:57	Success
CRS-10	C112	2017 Feb 19	First launch from KSC LC-39A since STS-135 in mid-2011. Berthing delayed by a day due to software problems.	23:08:08	Success
CRS-11	C106.2	2017 Jun 3	First mission to re-fly a recovered Dragon (previously flown on CRS-4).	27:01:53	Success
CRS-12	C113	2017 Aug 14	Last mission to use a new Dragon 1 spacecraft.	31:06:00	Success
CRS-13	C108.2	2017 Dec 15	Second reuse of Dragon. First NASA mission to fly aboard reused Falcon 9. First reuse of this Dragon.	25:21:21	Success
CRS-14	C110.2	2018 Apr 2	Third reuse of a Dragon. First reuse of this Dragon.	30:16:00	Success
CRS-15	C111.2	2018 Jun 29	Fourth reuse. First reuse of this specific Dragon spacecraft.	32:45:00	Success
CRS-16	C112.2	2018 Dec 5	Fifth reuse. First reuse of this Dragon. First-stage booster landing failed due to a grid fin hydraulic pump stall on reentry.	36:04:00	Success
CRS-17	C113.2	2019 May 4	Sixth reuse. First reuse of this specific Dragon spacecraft.	27:23:02	Success
CRS-18	C108.3	2019 Jul 24	Seventh reuse. First capsule to make a third flight.	30:20:24	Success
CRS-19	C106.3	2019 Dec 5	Eighth reuse. Second capsule to make a third flight.	29:19:54	Success
CRS-20	C112.3	2020 Mar 7	Ninth reuse. Third capsule to make third flight. Final launch of this Dragon version (Dragon 1).	28:22:12	Success

National Aeronautics and
Space Administration

OVERVIEW SpaceX CRS-21 Mission

SpaceX's 21st contracted cargo resupply mission (CRS) to the International Space Station for NASA will deliver more than 6,400 pounds of science and research, crew supplies and vehicle hardware to the orbital laboratory and its crew.

Launch is targeted for 11:39 a.m. EST Saturday, Dec. 5, 2020

Launch Vehicle
Falcon 9 Rocket

- Fourth flight of this booster
- Previous flights of this booster were Demo-2, ANASIS-II and a Starlink mission

Launch Site:
Launch Complex 39A,
NASA's Kennedy Space Center in Florida

Dragon Spacecraft Overview

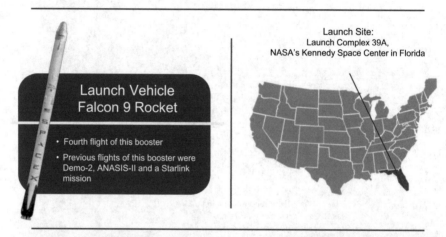

Height	8.1 m / 26.7 ft
Diameter	4 m / 13 ft
Capsule Volume	9.3 m³ / 328 ft³
Trunk Volume	37 m³ / 1300 ft³
Launch Payload Mass	6,000 kg / 12,228 lbs
Return Payload Mass	3,000 kg / 6,614 lbs

- First flight of a Dragon 2 Cargo Capsule, first Cargo Dragon to dock to the space station
- In January, it will re-enter Earth's atmosphere and splash down in the Atlantic Ocean near the eastern coast of Florida with 5,200 pounds of return cargo.

www.nasa.gov

For more information, visit www.nasa.gov/spacex

National Aeronautics and
Space Administration

CARGO

SpaceX CRS-21 Mission

*Masses are subject to change prior to launch

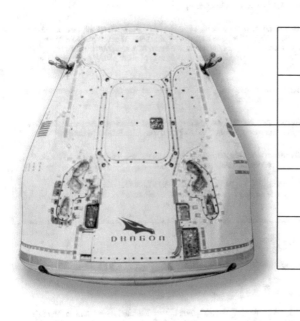

Crew Supplies
803 pounds / 364 kilograms

Science Investigations
2,100 pounds / 953 kilograms

Spacewalk Equipment
265 pounds / 120 kilograms

Vehicle Hardware
698 pounds / 317 kilograms

Computer Resources
102 pounds / 46 kilograms

Russian Hardware
53 pounds / 24 kilograms

Unpressurized Payloads
(Nanoracks Bishop Airlock)
2,403 pounds / 1,090 kilograms

Total Cargo:	**6,553 pounds / 2,972 kilograms**
Total Pressurized with Packaging:	**4,150 pounds / 1,882 kilograms**
Unpressurized Payloads:	**2,403 pounds / 1,090 kilograms**

www.nasa.gov

For more information, visit www.nasa.gov/spacex

National Aeronautics and
Space Administration

HARDWARE ▸ SpaceX CRS-21 Mission

Hardware Launching

Nanoracks Bishop Airlock and Installation Hardware: Bishop Airlock assembly with various installation support items to enable to crew to install the new airlock capability.

Exploration Catalytic Reactor: One of the main components of the Water Processor Assembly incorporating newly designed, robust metallic seals, a new catalyst, and oxygen flow regulation. These upgrades will help manage the dynamic temperature environment for a longer period and optimize the ability of the reactor to oxidize organic compounds in the water. Together, these modifications are intended to meet NASA's performance and reliability goals for a future Mars mission.

Nitrogen/Oxygen Recharge System (NORS) Recharge Tank: Supplemental nitrogen flying to support planned cabin repressurization activity aboard the space station.

Universal Waste Management System (UWMS) Spares/Consumables: Critical spares and consumable items to support crew usage of the next generation toilet following the four crew members arriving on Crew-1.

Rodent Research Habitats and Transporters: Live rodents and support hardware required for the Rodent Research-10 through-23 missions to be conducted during the Crew-1 timeframe.

One-handed Tape Dispenser: Through NASA's HUNCH challenge, high school students designed and fabricated a one-handed tape dispenser to provide astronauts an easily assessable tool for their everyday activities on the space station.

Hardware Returning

Failed or expended hardware no longer needed on the space station.

Treadmill Data Avionics Unit: Failed avionics unit that supports the treadmill, a critical item returning to the ground following the on-orbit replacement with a good spare.

Carbon Dioxide Removal Assembly (CDRA) Air Selector Valve: Critical degraded valve returning for repair and refurbishment to support the carbon dioxide removal capability on-orbit.

NORS Recharge Tank: Depressurized tank capable of flying oxygen or nitrogen, and will be utilized for future on-orbit demand in 2021.

Rodent Research Habitats and Transporters: Live rodents from the Rodent Research-23 mission and used habitats and transporters that support future research missions and analysis.

Minus Eighty Laboratory Freezer for ISS (MELFI) Electronics Unit: Failed cold stowage item requiring ground repair to enable future cold stowage missions.

Thermal Amine Bulk Water Save Valve: Failed valve that supports efficient usage of the Thermal Amine system returning to ground for repair, will help inform robustness of similar valve design on Orion.

National Aeronautics and
Space Administration

RESEARCH SpaceX CRS-21 Mission

The SpaceX cargo spacecraft will deliver dozens of investigations to the International Space Station, including research on microgravity effects on heart tissue, using microbes for mining on asteroids, how brain cells and tissues are affected by microgravity, how liquid metals behave in microgravity, a new privately funded airlock that can support science, CubeSat deployment, and spacewalks, and other cutting-edge investigations.

Certain microbes form layers on the surface of rock that can release metals and minerals, a process known as biomining. BioAsteroid examines biofilm formation and biomining of asteroid or meteorite material in microgravity. Researchers are seeking a better understanding of the basic physical processes that control these mixtures, such as gravity, convection, and mixing. Microbe-rock interactions have many potential uses in space exploration and off-Earth settlement. Microbes could break down rocks into soils for plant growth, for example, or extract elements useful for life support systems and production of medicines.

Microgravity causes changes in the workload and shape of the human heart, and it is still unknown if these changes could become permanent if a person lived more than a year in space. Cardinal Heart studies how changes in gravity affect cardiovascular cells at the cellular and tissue level. The investigation uses 3D engineered heart tissues (EHTs), a type of tissue chip. Results could provide new understanding of heart problems on Earth, help identify new treatments, and support development of screening measures to predict cardiovascular risk prior to spaceflight.

SUBSA-BRAINS examines differences in capillary flow, interface reactions, and bubble formation during the solidification of brazing alloys in microgravity. Brazing is a type of soldering used to bond together similar materials, such as an aluminum alloy to aluminum, or dissimilar ones such as aluminum alloy to ceramics, at high temperatures. The technology could serve as a tool for constructing human habitats and vehicles on future space missions as well as for repairing damage caused by micrometeoroids or space debris.

Launching in the trunk of the Dragon capsule, the Nanoracks Bishop Airlock is a commercial platform that can support a variety of scientific work on the space station. Its capabilities include deployment of free-flying payloads such as CubeSats and externally-mounted payloads, housing of small external payloads, jettisoning trash, and recovering external Orbital Replacement Units. Roughly five times larger than the airlock on the Japanese Experiment Module (JEM) already in use on the station, the Bishop Airlock allows robotic movement of more and larger packages to the exterior of the space station, including hardware to support spacewalks.

The Effect of Microgravity on Human Brain Organoids observes the response of brain organoids to microgravity. Organoids are small living masses of cells that interact and grow. They can survive for months, providing a model for understanding how cells and tissues adapt to environmental changes. Organoids grown from neurons or nerve cells exhibit normal processes such as responding to stimuli and stress. Therefore, organoids can be used to look at how microgravity affects survival, metabolism, and features of brain cells, including rudimentary cognitive function.

www.nasa.gov   For more information, visit www.nasa.gov/spacex

Appendix III

Commercial Crew Development (CCDev) Round 2

Commercial Crew Development Round 2

NASA's Commercial Crew Program (CCP) is investing in multiple American companies that are designing and developing transportation capabilities to and from low Earth orbit and the International Space Station (ISS).

Through the development and certification processes, NASA is laying the foundation for future commercial transportation capabilities. Ultimately, the goal is to lead to safe, reliable, affordable and more routine access to space so that commercial partners can market transportation services to the U.S. government and other customers.

After a transportation capability is certified, NASA would be able to purchase transportation services to meet its ISS crew rotation and emergency return obligations.

Through Commercial Crew Development Round 2 (CCDev2), NASA awarded $270 million in 2011 for the development of commercial rockets and spacecraft. This development round will be completed in mid- to late-2012.

The industry partners with whom NASA signed funded Space Act Agreements (SAAs) are Blue Origin, The Boeing Co., Sierra Nevada Corp. and Space Exploration Technologies (SpaceX).

The agency also signed unfunded agreements to establish a framework of collaboration with additional aerospace companies. As part of those agreements, NASA is reviewing and providing expert feedback to Alliant Techsystems Inc. (ATK), United Launch Alliance (ULA) and Excalibur Almaz Inc. (EAI) on overall concepts and designs, systems requirements, launch vehicle compatibility, testing and integration plans, and operational and facilities plans.

To find out more about the beginning of a new era in space exploration and NASA's Commercial Crew Program, visit www.nasa.gov/commercialcrew.

ATK
Liberty

NASA INVESTMENT: Unfunded	
PROFILE: Solid rocket boosters, Ariane 5 core stage, Vulcain 2 engine	
CAPABILITY: 44,500 pounds to low Earth orbit	

NASA and Alliant Techsystems Inc. (ATK) of Promontory, Utah, signed a Space Act Agreement (SAA) in September 2011 for the company's Liberty Launch Vehicle. Under the unfunded agreement, NASA's Commercial Crew Program (CCP) and ATK are exchanging technical information to aid in the development of Liberty as a commercial crew transportation provider.

Liberty's design combines the company's solid rocket boosters (SRBs) as the first stage and Astrium's Ariane 5 core stage and Vulcain 2 engine as the upper stage. The Ariane 5 rocket motors would form a two-stage launch vehicle with a single engine per stage. Astrium is a subsidiary of the European space company EADS.

Liberty's five-segment solid rocket first stage is derived from the Space Shuttle Program's four-segment solid rocket boosters. The Liberty first stage booster is capable of producing 3.6 million pounds of thrust at liftoff, roughly the same power as 63 four-engine 747 jets taking off. The five-segment first stage stands 154 feet tall. Combined with a modified Ariane 5 core stage as its upper stage, Liberty would be capable of carrying up to 44,500 pounds to low Earth orbit. Service and crew modules would be integrated to the top of Liberty in order to carry cargo and astronauts.

Other Liberty team members include United Space Alliance (USA) of Houston for launch vehicle integration and ground operations support and L-3 Communications of Cincinnati for first-stage avionics.

For more on ATK and Liberty, visit **www.atk.com**.

Image courtesy of ATK

© Springer Nature Switzerland AG 2022
E. Seedhouse, *SpaceX*, Springer Praxis Books,
https://doi.org/10.1007/978-3-030-99181-4

NASAfacts

Boeing
CST-100

NASA and The Boeing Company of Houston signed a funded Space Act Agreement (SAA) in March 2011 for the company's CST-100. Under the $92.3 million agreement, NASA's Commercial Crew Program (CCP) and Boeing are working to advance the company's Commercial Space Transportation System (CST). Optional milestones also were approved, valued at $20.6 million.

The CST-100 is a reusable, capsule-shaped spacecraft designed to take up to seven people, or a combination of people and cargo, to low Earth orbit, including the International Space Station (ISS).

The company has designed its spacecraft to be compatible with a variety of expendable launch vehicles and has selected United Launch Alliance's Atlas V rocket for initial CST-100 test flights.

Boeing also partnered with Pratt and Whitney Rocketdyne of Hartford, Conn., for the development of the spacecraft's Launch Abort Engine (LAE). The LAE combines Attitude Control Propulsion System thrusters from heritage spaceflight programs with a Bantam abort engine design and storable propellant engineering capabilities.

For more on Boeing and the CST-100, visit **www.boeing.com**.

NASA INVESTMENT: $92.3 million, plus $20.6 million in optional milestones
PROFILE: Crew module and service module
LANDING: Dry surface
CAPABILITY: Seven astronauts or equivalent crew and cargo

Image courtesy of Boeing

Blue Origin
Space Vehicle

Image courtesy of Blue Origin

NASA and Blue Origin of Kent, Wash., signed a funded Space Act Agreement (SAA) in March 2011 for the company's Space Vehicle. Under the $22 million agreement, NASA's Commercial Crew Program (CCP) and Blue Origin are working to advance the company's orbital concepts into a system capable of transporting four astronauts to low Earth orbit.

The Space Vehicle is designed to launch first on a United Launch Alliance Atlas V rocket and then on Blue Origin's reusable booster stage. The Space Vehicle would be capable of transporting four NASA crew members, wearing pressure suits, to the International Space Station. Missions would launch from Space Launch Complex-41 at Cape Canaveral Air Force Station, adjacent to NASA's Kennedy Space Center in Florida, and land at a Blue Origin facility in west Texas.

Blue Origin is performing wind tunnel testing at Lockheed Martin's High Speed Wind Tunnel Facility in Dallas to verify the spacecraft's aerodynamics. The company also is testing a pusher escape system that would rescue a crew if an emergency were to occur during launch or ascent, as well as the thrust chamber assembly (TCA) for the BE-3 100,000-pound thrust liquid oxygen, liquid hydrogen rocket engine that eventually would power the launch vehicle's reusable booster system.

For more on Blue Origin and the Space Vehicle, visit **www.blueorigin.com**.

NASA INVESTMENT: $22 million
PROFILE: Biconical spacecraft
LAUNCH: Atlas V rocket initially, then reusable booster stage
CAPABILITY: Four astronauts

Excalibur Almaz
Human Spacecraft

Image courtesy of Excalibur Almaz Inc.

NASA INVESTMENT: Unfunded
PROFILE: Human Spacecraft, which includes a capsule, launch abort system and expendable service module
LANDING: Ground landing
CAPABILITY: Three astronauts and/or cargo

NASA and Excalibur Almaz Inc. (EAI) of Houston signed a Space Act Agreement (SAA) in October 2011 for the company's Human Spacecraft. Under the unfunded agreement, NASA's Commercial Crew Program (CCP) and EAI are working to refurbish and upgrade the company's existing capsules with modern flight capabilities to launch humans to low Earth orbit.

During the partnership, NASA and EAI are conducting reviews of the Human Spacecraft's overall design, systems requirements, launch vehicle compatibility, testing plans, and operational and facilities plans. The company's heritage capsules, tested decades earlier, now will be upgraded with new internal systems and a service module to accommodate three crewmates during trips to low Earth orbit.

Astrium is providing EAI with propulsion subsystems, guidance and navigation and control components. Paragon Space Development Corp. of Houston is providing the Environmental Control and Life Support Systems (ECLSS). Lockheed Martin of Houston is performing systems engineering and integration while United Space Alliance of Houston is providing flight and ground operations support.

For more on Excalibur Almaz Inc. and the Human Spacecraft, visit **www.excaliburalmazinc.com**.

Sierra Nevada
Dream Chaser

NASA and Sierra Nevada Corp. (SNC) of Louisville, Colo., signed a funded Space Act Agreement (SAA) in March 2011 for the company's Dream Chaser spacecraft. Under the $80 million agreement, NASA's Commercial Crew Program (CCP) and Sierra Nevada are working to advance the company's reusable lifting-body spacecraft. Optional milestones also were approved, valued at $26.5 million.

The Dream Chaser is derived from NASA's HL-20, which somewhat resembles NASA's space shuttles and boasts years of development, analysis and wind tunnel testing by the agency's Langley Research Center in Hampton, Va.

Plans for the spacecraft include launching vertically and free-flight capabilities in low Earth orbit to dock with the International Space Station. Dream Chaser currently is the only Commercial Crew Development Round 2 (CCDev2) vehicle being developed with wings and the ability to land on a conventional runway.

Under the SAA, the company is developing non-toxic and storable propellants and assessing the vehicle's flight control surfaces and associated mechanisms required for flight.

Milestones under the agreement include a captive-carry completed May 29 and free-flight test of a full-scale prototype to test the vehicle's

Image courtesy of Sierra Nevada Corp.

approach and landing performance. A simulator consisting of a physical cockpit layout and integrated simulation hardware and software also will assist Dream Chaser engineers in evaluating the vehicle's characteristics during the piloted phases of flight.

The all-composite structure was designed by the SNC team and built in conjunction with SNC Dream Chaser team organizations AdamWorks of Centennial, Colo., Applied Composite Technology of Gunnison, Utah, and Scaled Composites of Mojave, Calif.

For more on Sierra Nevada and Dream Chaser, visit **www.sncorp.com**.

NASA INVESTMENT: $80 million, plus $25.6 million in optional milestones
PROFILE: Piloted lifting-body spacecraft
LANDING: Runway
CAPABILITY: Up to seven astronauts and cargo

SpaceX
Dragon

NASA and Space Exploration Technologies (SpaceX), of Hawthorne, Calif., signed a funded Space Act Agreement (SAA) in April 2011 for the company's Dragon capsule. Under the $75 million agreement, NASA's Commercial Crew Program (CCP) and SpaceX are working to outfit Dragon with life support systems and a launch abort system.

The reusable spacecraft is designed to launch atop the company's Falcon 9 rocket and would be capable of carrying up to seven astronauts to low Earth orbit. After a mission, the spacecraft would return to Earth's atmosphere with a parachute landing in the ocean. The company also is working on ground landing capabilities for the future.

The Dragon capsule currently is contracted to fly cargo-only missions to the space station for NASA's Commercial Resupply Services (CRS) Program. In 2010, the capsule became the first commercially developed spacecraft to return from Earth's orbit during a demonstration flight for the agency's Commercial Orbital Transportation Services (COTS) program.

Under the SAA, SpaceX is performing crew accommodation checks of its capsule and testing its SuperDraco engines. Eight SuperDracos would be built into the sidewalls of Dragon to carry astronauts to safety should an emergency occur during launch or ascent.

For more on SpaceX and Dragon, visit **www.spacex.com**.

NASA INVESTMENT: $75 million
PROFILE: Free-flying, reusable spacecraft
LANDING: Ocean initially, then ground
CAPABILITY: Seven astronauts

Image courtesy of SpaceX

ULA
Atlas V

NASA and United Launch Alliance (ULA) of Centennial, Colo., signed a Space Act Agreement (SAA) in July 2011 regarding the company's Atlas V rocket. Under the unfunded agreement, ULA is sharing its work to human-rate the Atlas V with NASA's Commercial Crew Program (CCP). The agency already relies on the Atlas V to launch complex scientific and robotic missions to space.

ULA is giving NASA an extensive look into its safety-critical launch vehicle systems, including the details of failure modes and effects analyses data from previous NASA missions, such as New Horizons, Juno and the Mars Science Laboratory.

Three of the four funded Commercial Crew Development Round 2 (CCDev2) partners have selected Atlas V as their launch vehicle including Sierra Nevada with its Dream Chaser, The Boeing Co. with its CST-100 spacecraft and Blue Origin for its Space Vehicle.

For more on ULA and the Atlas V, visit **www.ulalaunch.com**.

Image courtesy of United Launch Alliance

NASA INVESTMENT: Unfunded
PROFILE: Atlas core stage, Centaur upper stage and option for up to three solid rocket boosters
PROPULSION: More than 860,000 pounds of thrust from core stage at liftoff
CAPABILITY: Flexibility to launch different crew spacecraft

National Aeronautics and Space Administration

John F. Kennedy Space Center
Kennedy Space Center, FL 32899

www.nasa.gov

NASA Facts

FS-2012-07-125-KSC

Appendix IV

NASA's Commercial Crew Program

NASA's Commercial Crew Program

Goal

NASA's Commercial Crew Program (CCP) was formed to facilitate the development of a U.S. commercial crew space transportation capability with the goal of achieving safe, reliable and cost-effective access to and from the International Space Station and low-Earth orbit.

Background

CCP has invested in multiple American companies that are designing and developing transportation capabilities to and from low-Earth orbit and the International Space Station. By supporting the development of human spaceflight capabilities, NASA is laying the foundation for future commercial transportation capabilities.

Ultimately, the goal is to establish safe, reliable and cost-effective access to space. Once a transportation capability is certified to meet NASA requirements, the agency will fly missions to meet its space station crew rotation and emergency return obligations.

Throughout the process, both NASA and industry have invested time, money and resources in the development of their systems. NASA also is spurring economic growth through this program as potential new space markets are created.

To accelerate the program's efforts and reduce the gap in American human spaceflight capabilities, NASA awarded more than $8.2 billion in Space Act Agreements (SAAs) and contracts under two Commercial Crew Development (CCDev) phases, the Commercial Crew Integrated Capability (CCiCap) initiative, Certification Products Contract (CPC) and Commercial Crew Transportation Capability (CCtCap).

CCP is primarily based at NASA's Kennedy Space Center in Florida, the space agency's premier launch site. About 200 people are working in CCP for NASA, with almost half involved in the work at other NASA centers, including Johnson Space Center in Houston and Marshall Space Flight Center in Huntsville, Alabama.

© Springer Nature Switzerland AG 2022
E. Seedhouse, *SpaceX*, Springer Praxis Books,
https://doi.org/10.1007/978-3-030-99181-4

NASAfacts

How NASA's Commercial Crew Program is Different

naSa's prior approach for Obtaining Crew transportation SyStemS:
- NASA devised requirements for a crew transportation system that would carry astronauts into orbit, then the agency's engineers and specialists oversaw every development aspect of the spacecraft, its support systems and operations plans.
- An aerospace contractor was hired to build the crew transportation system to the design criteria and the standards NASA furnished.
- NASA personnel were deeply involved in the processing, testing, launching and operation of the crew transportation system to ensure safety and reliability. The space agency owned the spacecraft and its operating infrastructure.
- Every spacecraft built for humans, from Mercury to Gemini and Apollo to the space shuttle and American section of the International Space Station, was built and operated using this model.

Commercial Crew's approach for Obtaining Crew transportation SyStemS:
- NASA's engineers and aerospace specialists work closely with companies to develop crew transportation systems that can safely, reliably and cost-effectively carry humans to low-Earth orbit, including the International Space Station, and return safely to Earth.
- Interested companies are free to design the transportation system they think is best. For the contract phases of development and certification, each company must meet NASA's pre-determined set of requirements.
- The companies are encouraged to apply their most efficient and effective manufacturing and business operating techniques throughout the process.
- The companies own and operate their own spacecraft and infrastructure.
- The partnership approach allows NASA engineers insight into a company's development process while the agency's technical expertise and resources are accessible to a company.

Complementary Approaches to Complementary Goals

Space act agreements: When NASA decided to support the development of new U.S. human spaceflight capabilities to low-Earth orbit, it relied on its commercial partners to propose specifics, ranging from the design and capabilities to private investment ratio, milestone achievements, success criteria and timelines. Once an agreement was accepted, CCP and its expert teams worked closely with each company to provide technical support and determine when milestones were met.

ContractS: Concurrently with the Space Act Agreements, NASA established safety and mission requirements for missions to the International Space Station that would be flown under a NASA contract. During their development efforts, companies could choose to design their systems to meet NASA's pre-determined requirements. To support the certification of these systems, NASA awarded Certification Products Contract (CPC) and Commercial Crew Transportation Capability (CCtCap) contracts.

Commercial Development with Space Act Agreements

To support the goal of achieving safe, reliable and cost-effective access to and from low-Earth orbit for commercial customers, NASA used Space Act Agreements to partner with domestic companies capable of contributing to the development of a U.S. human spaceflight capability.

Commercial Crew Development round 1 (ccDev1)
Space Act Agreement
As NASA retired the space shuttle, the ability of private industry to take on the task of providing routine access to space was of vital importance. In 2010, NASA invested a total of nearly $50 million of the American Recovery and Reinvestment Act (ARRA) funds for CCDev1 to stimulate efforts within the private sector to aid in the development and demonstration of safe, reliable and cost-effective crew transportation capabilities. It included the development and maturation systems and subsystems, such as a spacecraft, launch vehicle, launch abort systems, environmental control and life support system, launch vehicle emergency detection systems and more.

> Blue Origin - $3.7 million
> Boeing - $18 million
> Paragon Space Development Corporation - $1.4 million
> Sierra Nevada Corporation (SNC) - $20 million
> United Launch Alliance (ULA) - $6.7 million

Commercial Crew Development round 2 (ccDev2)
Space Act Agreement
CCDev2 kicked off in April 2011 when NASA awarded a total of nearly $270 million to four companies to aid in further development and demonstration of safe, reliable and cost-effective transportation capabilities. The agency also signed unfunded Space Act Agreements to establish a framework of collaboration with additional aerospace companies.

NASA Facts

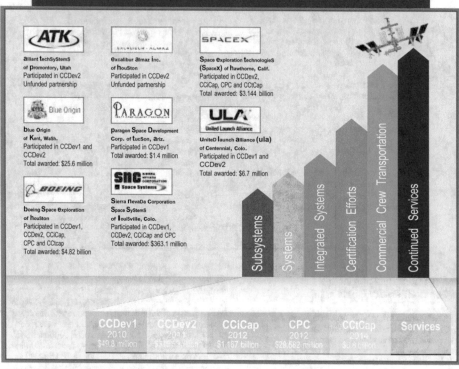

As part of those agreements, NASA reviewed and provided expert feedback on overall concepts and designs, systems requirements, launch vehicle compatibility, testing and integration plans, and operational and facilities plans.

> Alliant Techsystems Inc. (ATK) - unfunded
> Blue Origin - $22 million
> Boeing - $92.3 million
> Excalibur Almaz Inc. (EAI) - unfunded
> Sierra Nevada Corporation (SNC) - $80 million
> SpaceX - $75 million
> United Launch Alliance (ULA) – unfunded

NASA later funded an additional $20.6 million to Boeing and $25.6 million to Sierra Nevada Corporation by exercising optional, pre-negotiated milestones, which were part of their original Space Act Agreements, to accelerate development.

In 2012, the agency extended its CCDev2 agreement with Blue Origin in an unfunded capacity. Through the agreement, the agency continued to support the development of the company's Space Vehicle and related systems.

Commercial Crew IntegrateD Capability (CCiCap)
Space Act Agreement
CCiCap continued the development of three fully integrated systems. The Space Act Agreements called for industry partners to develop crew transportation capabilities and to perform tests to verify, validate and mature integrated designs.

> Boeing - $460 million
> Sierra Nevada Corporation (SNC) - $212.5 million
> SpaceX - $440 million

NASA later funded an additional $20 million to Boeing, $20 million to SpaceX and $15 million to Sierra Nevada Corporation by exercising optional, pre-negotiated milestones, which were part of their original Space Act Agreements, to accelerate development.

NASA Facts

Supporting NASA's Mission Needs through Contracts

To support the goal of achieving safe, reliable and cost-effective access to and from the International Space Station for the agency, NASA awarded contracts intended to permit the certification of commercial crew transportation systems to carry NASA astronauts.

Certification ProDuctS ContractS (cpc) Contract

Throughout CPC, the first phase of a two-phase contract, companies worked with NASA to discuss and develop data products to implement the agency's flight safety and performance requirements. This included implementation across all aspects of the space system, including the spacecraft, launch vehicle, and ground and mission operations.

Under the contracts, certification plans were developed toward achieving safe, crewed missions to the space station. It included data that will result in developing engineering standards, tests and analyses of crew transportation system designs. NASA awarded a total of nearly $30 million under the CPC contracts.

> Boeing - $9.993 million
> Sierra Nevada Corporation (SNC) - $10 million
> SpaceX - $9.589 million

Commercial Crew tranSportation Capability (CCtCap) Contract

CCtCap is the second phase of a two-phase certification plan for commercially built and operated integrated crew transportation systems. Two FAR-based, firm fixed-price contracts were awarded in September 2014 following an open competition. Through its certification efforts, NASA will ensure the selected commercial transportation systems meet the agency's safety and performance requirements for transporting NASA crews to the International Space Station. NASA awarded a total of $6.8 billion under CCtCap contracts.

> Boeing - $4.2 billion
> SpaceX - $2.6 billion

National Aeronautics and Space Administration

John F. Kennedy Space Center
Kennedy Space Center, FL 32899

www.nasa.gov

NASA Facts

FS-2014-010-284-KSC

Index

© Springer Nature Switzerland AG 2022
E. Seedhouse, *SpaceX*, Springer Praxis Books,
https://doi.org/10.1007/978-3-030-99181-4

Printed in the United States
by Baker & Taylor Publisher Services